# AMERICAN FARM TOOLS

## from Hand-Power to Steam-Power

R. Douglas Hurt

ISBN 0-89745-026-4 paper
ISBN 0-89745-027-2 hardback

From the January 1982 *Journal of the West.*



Third Printing, 1985

Sunflower University Press®
1531 YUMA (BOX 1009) • MANHATTAN, KANSAS 66502, USA

# AMERICAN
# FARM TOOLS

## from
## Hand-Power to Steam-Power

**R. Douglas Hurt**

# *PREFACE*

Today, only about 3 percent of the American population live on farms. Yet, many people still have an interest in agriculture and are eager to learn about their nation's rural heritage. Still, while agricultural tools and implements can be studied at several major technological museums, state and local historical societies, or living historical farms, no single museum presents a comprehensive collection of American agricultural technology for public viewing. Furthermore, farming methods are altered over time as changes are made in agricultural science and technology. Consequently, many tools and implements that were once commonly used and understood by the majority of Americans are now only vaguely recognized, if they are remembered or identified at all. For young Americans who assume that milk originated in plastic bottles or that hamburger is produced in fast food restaurants, the agricultural past is completely alien.

Because of the interest of many and the need of some to learn more about farming, my intent is to provide a basic reference about technological change in American agriculture. Specifically, my purpose is to trace technological change from the days of hand power through the steam age, or approximately from the colonial period to the beginning of the First World War. In order to define an almost boundless subject, I have focused on the basic tools and implements used for the production of cereal grains, hay and fodder. I have broadly defined farm tools to include not only hand held and manipulated objects, but also implements and machines which enable the farmer to produce more efficiently and more abundantly than he possibly could without them.

Any endeavor to cover such a broad expanse of time, of course, presents many problems, the most serious of which is to provide a comprehensive survey without becoming either overly detailed or superficial. Space limitations for the number of tools and implements that can be illustrated also present problems. Consequently, I have not attempted to discuss or list every implement invented or used relating to the topics of discussion. Any attempt to do so would be impossible. Nor have I tried to identify every inventor who first patented the specific agricultural innovation mentioned in the text. Inventors were numerous, and the time between the patenting of an implement and its perfection, manufacture, and practical application was usually long. Consequently, it is frequently impossible to attribute particular developments to specific inventors. Rather, my purpose is to provide a chronological discussion of the technological developments which changed the nature of farming and which stimulated further innovation of American agricultural technology.

* * *

I am grateful to Homer E. Socolofsky and to Chris Duckworth for reading all or portions of the manuscript, and to John T. Schlebecker who has influenced my thoughts about technological change in American agriculture. Bob Walther at the Smithsonian Institution's National Museum of American History generously aided with the selection of the illustrations. His help was beneficial beyond proper recognition. Hiram M. Drache provided several of the photographs for which I am indebted. I am thankful as well for the help which I received at the libraries of the Ohio Historical Society and Ohio State University, and from the staff members in the Still Pictures Branch at the National Archives. I am particularly grateful to Mary Ellen Hurt for preparing the index. Lastly, I am indebted to William G. Keener, Associate Director of the Ohio Historical Society, for making this study possible.

*Farm tools in use in 1790. (Smithsonian Institution.)*

# *CHAPTER I*

# *INTRODUCTION*

Between the founding of Jamestown in 1607 and the beginning of the First World War in 1914, the tools and implements used on the American farm underwent vast change. Whether that technological change was revolutionary or evolutionary depends, of course, on one's perspective and interpretation of the past. No one can doubt, however, that, over time, technological invention profoundly influenced the farmer's way of life. Moreover, technological innovation gave a particular shape to the American past. It is the shape of time and the shape of change, and it provides an historical record that can be studied as profitably as any written document.

Technological change in American agriculture, however, did not just happen. No one simply decided to build a mechanical reaper in order to harvest larger acreages more quickly than one could possibly cut with a cradle scythe. No one simply invented a machine to thresh grain in order to dispense with the flail and winnowing basket. Technological change takes time, and it depends upon three criteria. First, it requires cumulative knowledge. Before anyone could build a steam engine someone had to invent the wheel. Success in technological innovation depends upon knowledge gained from prior experience. An inventor draws upon the past — accepting, rejecting, and synthesizing — to shape a new idea into a workable product. Secondly, technological change relies upon a perceived need. In agriculture, as well as in other endeavors, the new invention must clearly work to the owner's advantage. If nothing is to be gained, such as plowing more easily, reaping more quickly, or threshing more efficiently, there would be no reason to adopt the new invention. Third, the product of technological innovation must be affordable. If farmers had not had sufficient means to purchase a cast-iron plow, that invention would have never replaced the wooden moldboard.

During the colonial age, the tools which the farmer used were little different from those employed by the farmers in the Roman Empire. In colonial America, the farmer or the local blacksmith fashioned most of the required tools; or, if the farmer was wealthy, he might purchase some of the needed tools from Europe. More often than not, however, farm tools were fashioned at home after British models, but those tools were usually modified to meet local conditions.

From the colonial to the early national period of American history, the basic farm tools were the hoe, spade, and plow for tillage; the sickle, scythe, and hand rake for harvesting and mowing; and, the flail and winnowing basket for threshing. Between 1790 and1865, however, American agriculture underwent rapid change. Some implements, such as the plow, were improved in design, and new methods of manufacture were developed. Iron replaced wood on many implements, and inventors applied the concept of interchangeable parts to farm tools. Some innovations produced new technological forms such as mowing machines, sulky plows and self-rake reapers. In addition, power sources changed from hand to horse to steam. Many inventions involved providing special solutions for particular problems. Quickly, these specific solutions became general solutions as in the case of the development of a steel plow for breaking western prairie sod. Soon, farmers in all regions wanted steel plows for their tillage operations. Some inventions simply enabled the farmer to perform necessary tasks more efficiently than ever before, for example, the substitution of the threshing machine for the flail.

Most inventors did not build or manufacture their tools and implements for immediate sale. If they lacked capital, access to skilled labor, and shop facilities, they sold licenses to others who had those resources. Once a new tool became reasonably effective, farmers began to use it on a limited scale. Most farmers, however, were reluctant to purchase implements that had not been thoroughly proven. Others were hesitant to invest in new tools and implements, if their lands were unsuited for the application of that technology. Rocky soil in New England and stump-filled clearings in Kentucky made them wait until either the rocks and stumps were removed or until the technology could accomodate those special conditions. This hesitation quickly faded during the Civil War, when labor shortages, high prices, and wartime demands encouraged farmers to invest in the new technology in order to produce larger crops and to reap more substantial profits than ever before. When the war ended, further technological innovations continued to stimulate farmers to adopt additional tools and implements.

Farmers, who were either unwilling or reluctant to adopt the new technology, were coaxed continually into doing so by implement company advertisements in agricultural periodicals and newspapers, by equipment exhibits at agricultural fairs, and by demonstrations of traveling salesmen. Agricultural societies, the United States Department of Agriculture, and the land-grant colleges also disseminated information on technological changes that would ease the farmer's burdens and improve the efficiency and profits of the farm operation. Usually, if an implement saved time, eased toil, expanded production, and decreased costs, farmers were inclined to adopt it. Generally, they tended to purchase the most indispensable implement at the cheapest possible price.

Technological advance in one area, however, required comparable change in other areas. If, for example, a farmer purchased a grain drill to seed more wheat,

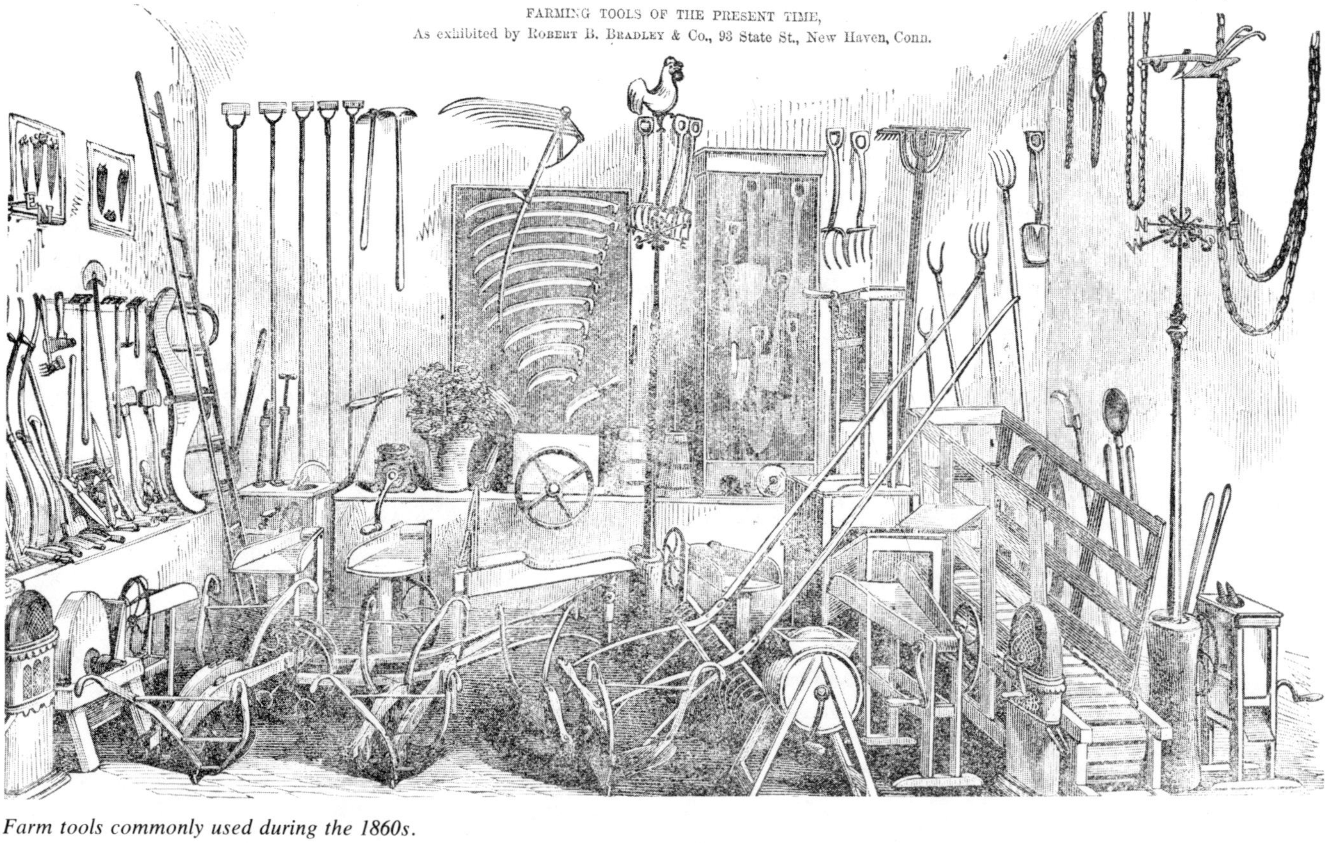


*Farm tools commonly used during the 1860s.*

this implement could not benefit him unless he also had the technology to reap a larger crop. If the crop shriveled from heat or was destroyed by grasshoppers or by foul weather before the harvest could be completed, the grain drill did not provide the farmer any significant advantage. Technological balance, however, had been largely achieved by the mid-nineteenth century.

By the turn of the twentieth century, technological change on the American farm had been phenomenal. In 1900, someone born at the end of the War of 1812 could have been able to recount harvesting wheat with a sickle, cradle, reaper, and binder. One could recall threshing the crop with a flail and separator as well as turning the furrow with wooden, steel and sulky plows. At eighty-five years of age, someone could remember cutting hay with a scythe and mower and using a wooden horse-rake and steel side-delivery rake to put it into a windrow. Such a person might have watched a combine operate or even used one. Certainly, one could give instruction about how a steam engine worked. Indeed, no other aspect of American life was more profoundly affected by technological change prior to the perfection of the internal combustion engine than was agriculture. Technological change in agriculture contributed to the nation's economic growth, released farm workers for industry, and enabled the production of an abundant food supply for urban America. No other form of technological change would affect American life more significantly than that applied to agriculture prior to the rapid development of the automobile industry during the 1920's.

## *AUTHOR'S NOTE*

Several years ago, Robin Higham asked me to write a piece about farm tools in the American West. His request excited and challenged me. Almost immediately, however, I confronted the problem every historian must contend with when writing about the West. Simply put, Where is the West? Does it begin at the Mississippi River, the Great Plains or the Rocky Mountains. Or, is it the line of demarcation stretching from Saint Paul to Fort Worth which Gilbert Fite used for his seminal study *The Farmers' Frontier, 1865-1900*. All of these boundaries for marking where the West begins seemed reasonable, but, at the same time, each was not quite satisfactory for this study.

Confronted with a problem of definition, I decided to avoid it for the moment and to concentrate my research on western farm tools anyway. Again, I was immediately confronted with another equally serious problem. It was that very few farm tools, used within the scope of this study, were developed entirely in the West. Virtually every tool had eastern antecedents. Certainly, some tools were better suited for western conditions than were others, and, ultimately, some tools were perfected in the West. Still, these factors did not make them specifically *western* farm tools. So, my problems compounded.

Finally, the idea struck me (though hardly with a bolt of lightning), that there has always been a West. In this sense, the West is a state of mind. This revelation will hardly be new or astounding for historians of the American West, but it did solve my problems. Certainly, from the time the first Europeans settled the North American Continent, the West has always been those lands which lay beyond. If not, I am badly mistaken.

With this rational in mind, my work proceeded smoothly. For those readers, however, who do not accept this reasoning, I must apologize and urge them to try again.

R. Douglas Hurt earned his Ph.D. in American History at Kansas State University. He has been a Smithsonian Fellow in the History of Science and Technology, and he has taught at Texas Tech and Ohio State Universities. His specialization is agricultural history, and he is the author of *The Dust Bowl: An Agricultural and Social History* (1981) as well as a number of articles for scholarly journals. He is Curator of Agriculture at the Ohio Historical Society.

*Through the ages, the plow has been the most important tool for tilling the soil. This New England farmer is using oxen to pull an iron-beamed walking or swing plow. (George E. Tingley photo, 1899. Library of Congress.)*

# CHAPTER II

# THE PLOWMAN'S TOOLS

## PLOWS

Through the ages the plow has been the most important agricultural tool. Indeed, without it farmers could not till the soil and prepare their fields for extensive agriculture. Although the plow is an ancient agricultural tool, it underwent little change prior to the invention of improved methods for making iron and steel in the mid-nineteenth century. American plowmen, however, always demanded an implement which required little draft, that is, the amount of power needed for pulling it, as well as one which ran at a uniform depth, turned over the furrow, and pulverized the soil. But, these demands were seldom met before the standardization of design and the perfection of interchangeable parts.

During the early seventeenth century, though, if farmers owned plows at all, they were usually British imports or were crudely fashioned homemade tools. Most beginning colonial farmers went without plows for a considerable period of time. Instead, they used hoes and mattocks to prepare the seedbed. The Pilgrims, for example, did not have plows until 1632 — 12 years after their arrival at Plymouth Rock. Only a few miles to the north, in 1637, the Puritan farmers around Boston had only 37 plows to till their fields. And, as late as 1642, Rhode Island farmers were still using hoes and spades to turn the soil. Later, British mercantile policy intentionally restricted the development of American industry to make the colonists reliant upon England for manufactured goods. Consequently, colonial farmers could either import expensive English plows or fashion their own as best they could.

Because of the general absence of plows in colonial America, farmers who owned one tilled their neighbor's fields. Or, the town paid a bounty to any farmer who purchased a plow and used it to prepare local fields for planting. Those farmers, who could neither afford a plow nor hire their plowing done, fashioned plows of their own design from the wood and metal available. They commonly selected a winding tree and crafted a moldboard from it. Ideally, the moldboard's function was to lift the furrow slice, turn it over, and bury the crop stubble. This procedure would leave the plowed field relatively smooth and in a suitable condition for further tilling and planting. In order to prevent the

*Colonial farmers in New England used the Old Colony Strong Plow to break virgin soil. The wooden moldboard was plated with thin iron straps to reduce friction and to prevent the soil from sticking to the moldboard. The iron bar in front of the moldboard is called a coulter. The coulter cuts the furrow slice vertically while the moldboard turns it over horizontally. Eastern farmers continued to use the Old Colony Plow until about 1820. It had a 10-foot beam and a 4-foot landside (the side of the plow which rested against the soil when the moldboard turned the furrow). This plow reportedly made furrows stand up like the "ribs of a lean horse in the month of March." (Smithsonian Institution.)*

moldboard from wearing out too rapidly, farmers plated it with iron from worn out saw blades, hoes, and horse shoes. Before this iron could be attached to the moldboard, however, the local blacksmith heated and pounded it into thin strips which were then nailed or bolted onto the face of the plow. The side opposite the moldboard, called the landside, was made from an iron bar or strap, and the bottom of the plow was shod with a thin iron plate. The share or cutting edge of the plow was also made from iron. The beam and handles were wooden and were fashioned respectively from a tree trunk and from crooked branches. All of these parts were attached in a somewhat haphazard fashion. The six- or seven-foot beam was set at any pitch the farmer desired, and the handles were usually fastened at nearly right angles — both practices of which gave the farmer very little control over his implement. Two or three yoke of oxen were required to pull it. Even so, tilling the soil with plows such as these required an extraordinary amount of hard work, and the process was invariably slow, since the cutting and turning ability of these implements was undependable. The net result was an imperfectly prepared seedbed.

When the farmer's homemade plow broke, he had little hope of repairing it exactly. Or, if a farmer had purchased a particulary effective implement from a plowright, he still had no guarantee that the craftman's repairs would return the plow to its former level of performance. In short, if a farmer owned a plow which easily cut through the soil and turned the furrow smoothly and completely, it was because of accident rather than from design.

Thomas Jefferson, who was a farmer, thought the plow could be designed on mathematical principles so that a standardized moldboard could be easily reproduced. Such plows would provide maximum tilling ability and at the same time reduce the draft or power required for pulling the implement — all of which would make plowing easier for man and beast alike. At the same time, scientifically designed plows would do a better job of turning the furrow and killing weeds, since standardized moldboards could be fashioned for all soil types. In 1798, with these goals in mind, Jefferson designed a moldboard that would lift the soil vertically and, in a continuous motion, turn it over horizontally. This moldboard, however, did not turn all soils in a uniform manner, and it was never manufactured on a commercial basis. Nevertheless, Jefferson successfully demonstrated that a standardized moldboard could be produced, provided the mathematical formula was perfected for its design. And, while Jefferson made no attempt to improve the technical aspects of the plow's sole, landside, and position of the beam and share, he did recommend casting the moldboard from iron to improve the plow's cutting and wearing ability. This was an important idea, because standardization of design could not be achieved by using wood, since each plowmaker fashioned it as he pleased. Only metal which was cast, wrought or molded in some fashion would permit consistent duplication of superior design.

Jefferson never cast his moldboard, but, in 1797, Charles Newbold, a New Jersey inventor, patented his own plan for a cast-iron plow. Newbold cast the moldboard, share and landside in one, solid piece. Although Newbold proved that such casting could be done, it was far from practical. Indeed, if any part

*In 1797, Charles Newbold patented a cast iron plow. Newbold fashioned the moldboard, share, and landside in one solid piece. This illustration shows the Newbold plow from the landside. The Newbold plow was too expensive for the average farmer to afford and many farmers believed that cast-iron poisoned the soil.*

broke, as happened to the original model's share point during testing, the entire plow became useless. If the plow did not break, the share dulled quickly and the farmer had to either replace the entire implement or sharpen it frequently. These unfortunate features made the plow far too expensive for the average farmer. Many farmers also apparently believed cast-iron poisoned the soil, and encouraged the weeds to grow and, therefore, refused to adopt it for these reasons. Even though Newbold substituted a wrought-iron share, the American farmer still preferred the wooden moldboard. Still, Newbold's invention was a major advance in plow concept, design and construction. Other agricultural inventors would build on his technological contribution.

Several years later, in the spring of 1807, David Peacock, also from New Jersey, patented a cast-iron plow with three parts. The moldboard and the landside were cast separately and a wrought-iron, steel-edged share was attached. Peacock's design was more practical than Newbold's since a worn out or broken part could be replaced. About this same time, the prejudice against the cast-iron plow began to fade away and Peacock's plows gained widespread popularity in the middle Atlantic states where they were used until the eve of the Civil War.

The concept of standardized, replaceable parts is, however, usually credited to Jethro Wood of Scipio, New York. Probably because he was the most successful inventor to market a plow with these features. In 1814, when Wood patented his plow with replaceable parts, he probably knew about Peacock's design, be-

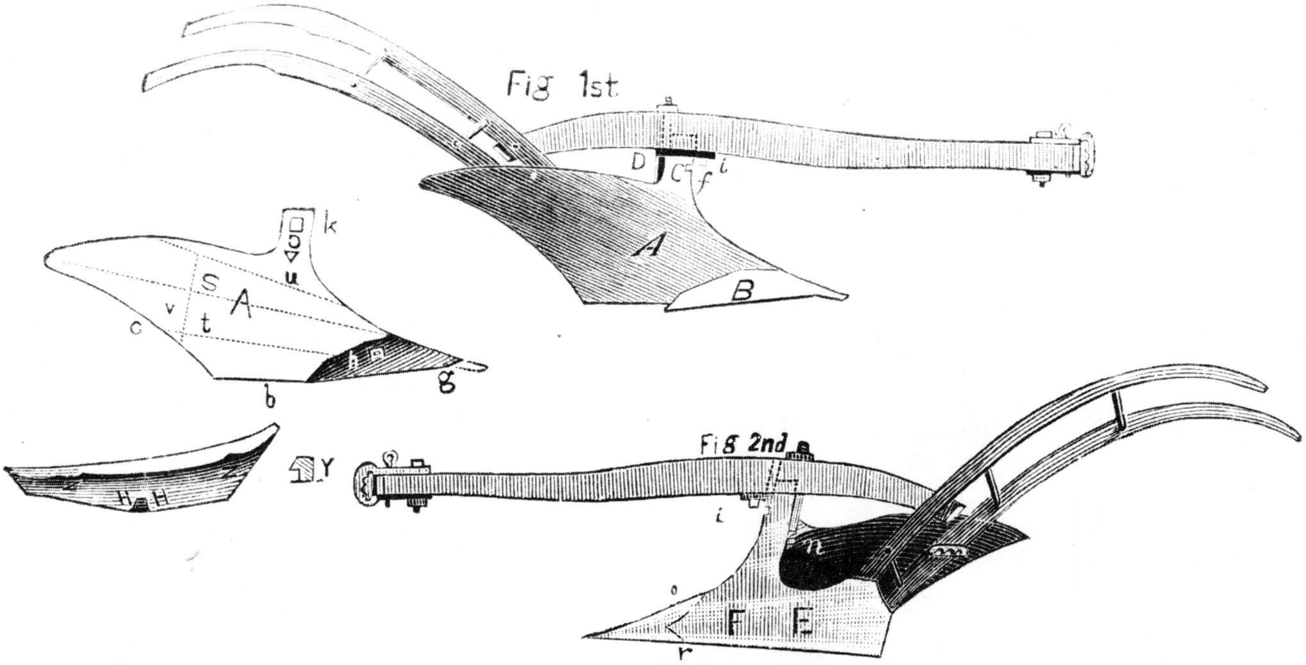


FAC-SIMILE OF THE ORIGINAL WOOD PLOW.

*In 1814, Jethro Wood patented an iron plow with interchangeable parts. Wood improved his design in 1819. The great advantage of Wood's plow was that if one part broke, it could be easily replaced. (Smithsonian Institution.)*

# SHOVEL PLOWS

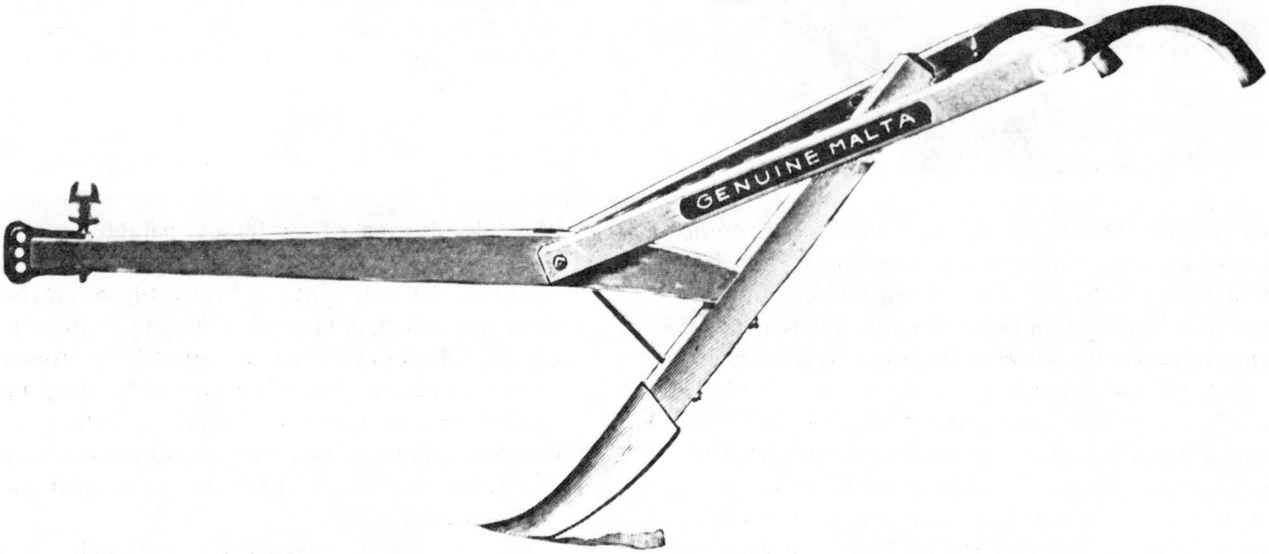


*The shovel plow could be made by attaching a wrought-iron blade, with the convex side turned outwards, to a beam. The shovel plow cut a shallow furrow, and it was popular in the South for tilling the soil. Northern farmers primarily used it for cultivating soil.*

cause he did not claim that he had invented the principle of interchangeable parts. In 1819, however, when he patented an improved design, he did claim a new method for joining the moldboard, landside and share without the use of screws or bolts which were difficult to make or expensive to obtain. Various mortise and tenon joints allowed the pieces to lock together. A steel-tipped share cut through the soil more efficiently and required less sharpening than cast-iron shares. Wood's plow probably did more to eliminate the old, clumsy, wooden plows than any other design to that time, and farmers were quick to purchase it. In 1817, more than 1,500 of Wood's plows were sold in New York. A year later, 1,600 plows were sold; and, in 1819, the year he patented his improved model, New York farmers purchased 3,600 of his plows. Wood's plow remained popular for decades and stimulated other inventors to fashion their own plow designs after his model. Most of the plows patented for a long time thereafter, differed very little in their general principles.

Even though an estimated 10,000 plows had been manufactured in the United States by 1820, the limited advantages of the cast-iron plow prevented farmers from quickly purchasing the implement on a wide basis. Some farmers were too conservative or reluctant to try these new implements since they had used nothing but wooden moldboard plows all of their lives. Most, however, simply could not afford to purchase a cast-iron plow. Cast-iron plows cost approximately twice as much as a wooden plow, if both were purchased from a merchant. If the farmer made his wooden plow at home he could save even more money. Furthermore, plow parts were not always readily available for the new cast-iron models, because of poor transportation and distribution systems in nineteenth century America. The benefits of interchangeable parts were of little value, if the parts could not be obtained in the first place. And, finally, although the cast-iron plows often turned the furrow more efficiently than wooden plows, they did not always perform as well as the best wooden models. Generally, these early cast-iron plows had difficulty penetrating the soil deeply and were able to turn only a four-or five-inch furrow. Consequently, even though cast-iron plows required substantially less draft, approximately one yoke of oxen instead of the two or three yoke needed for a wooden plow, many farmers took a wait-and-see attitude.

*The Carey plow was popular among farmers in the early nineteenth century. This plow had a wooden moldboard plated with iron strips. A wrought-iron share was attached in front of the moldboard to split the furrow. (Smithsonian Institution.)*

Although farmers increasingly adopted cast-iron plows between 1820 and 1835, the wooden moldboard

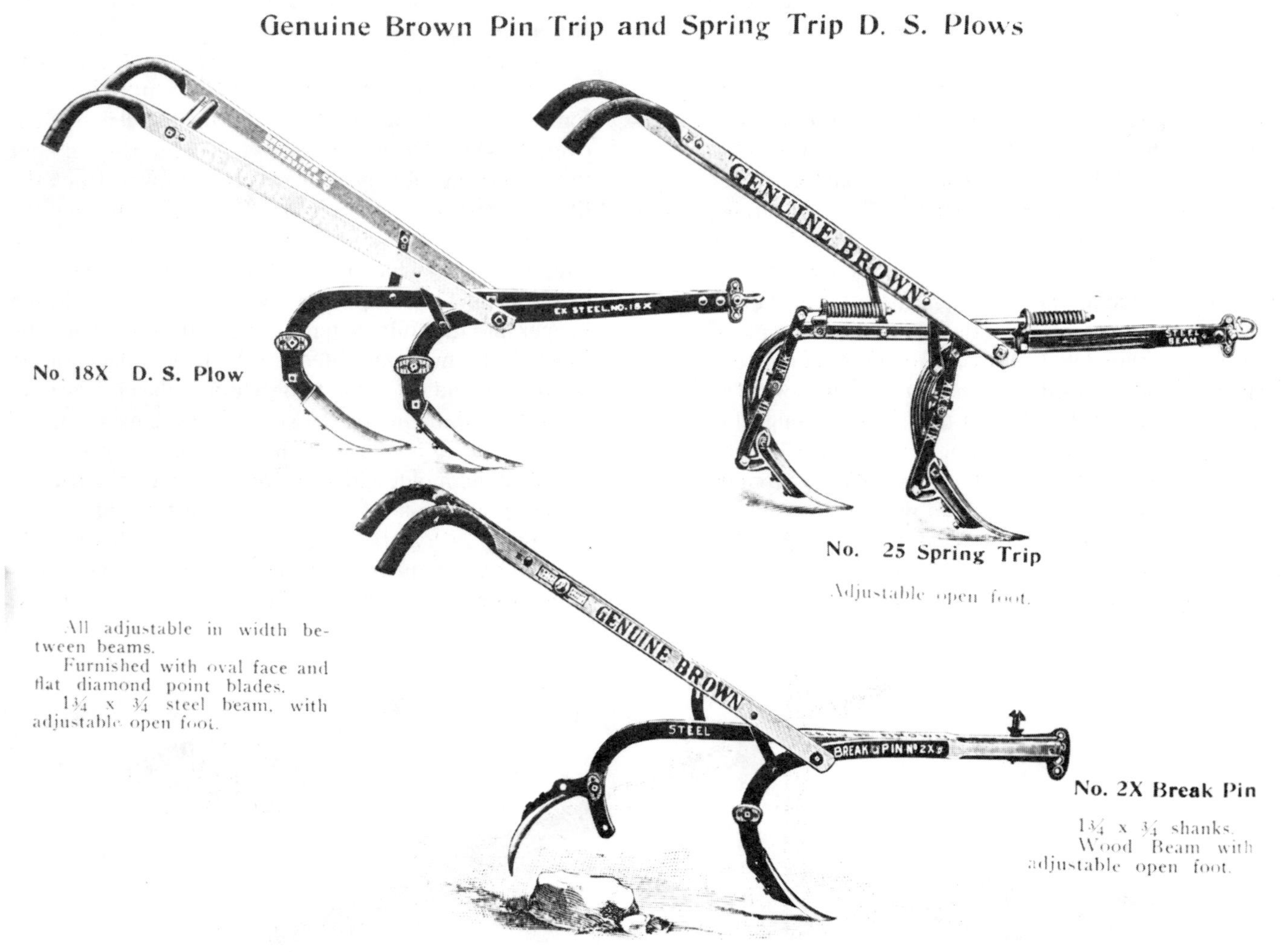


*Some shovel plows, such as these manufactured by the Brown-Manly Plow Company of Malta, Ohio, were fitted with wooden break pins or springs. When the shovel hit a solid obstruction, the pin broke or the spring flexed, thereby, allowing the shovel to bend backwards. This innovation helped prevent broken or damaged shovels.*

plow remained a favorite. The Carey plow was perhaps the most extensively used wooden model. Although the Carey plow's form differed somewhat according to the skill of each blacksmith or plowright who worked on it, the general style was uniformly reproduced on a wide basis. The Carey plow had a wooden landside and moldboard. Iron straps plated the moldboard and a wrought-iron share was attached to it. The beam and handles were also made from wood. All joints were wooden and the various pieces were attached with wooden pegs. Over time, these joints loosened and the wood cracked or broke, all of which made plowing difficult and repairs frequent. Still, the Carey plow was popular in the North as well as in the South. With it, a farmer could plow about one acre per day.

Many Southern farmers, however, preferred to use the shovel plow above all others primarily because they maintained their prejudice against the cast-iron plow. The shovel plow, which had become popular in the colonies prior to the American Revolution, was usually made by the plantation blacksmith. It consisted of a rough wooden beam into which another wooden piece with an iron point was attached. Two handles were nailed or pinned to the sides of the beam. The wrought-iron point, approximately nine inches across, was shaped much like a shovel with the convex side turned outwards. A loop was welded on the back of the blade to provide a place for the stock to enter. One horse or mule could easily pull the shovel plow. It cut a shallow furrow and turned the soil both ways. The nearly upright position of the handles forced the plowman to maintain an erect, tiring position. One observer noted that using a shovel plow was "pretty much like dragging a cat by the tail." Nevertheless, southern farmers continued to use it for plowing and cultivating until the Civil War, particularly in the coastal and piedmont regions of Georgia and the Carolinas. Nevertheless, while plantation profits were primarily invested in land and slaves rather than in agricultural implements, many southern farmers began using the Carey plow on an ever increasing basis.

Although plowrights fashioned moldboards into various shapes in order to meet the requirements of different soils, more adjustments were necessary, when farmers crossed the Appalachians and entered the rich prairie land of the Midwest. There, the tough prairie sod made the wooden plow useless. It would neither penetrate the sod and cut the roots nor stay in the soil. Consequently, prairie farmers used another plow to turn a furrow. That tool, called a breaking plow, was similar to the Old Colony Strong Plow which New England farmers used for breaking virgin soil. The prairie breaker, popular from the 1820s, through the 1840s was an immense, wooden plow plated with iron strips to reduce as much friction as possible. A wrought-iron share and coulter were fitted onto the moldboard and beam to cut through the tough root system of the grass. Some models had a long, sloping, iron moldboard, designed to lift the sod and turn it over completely in order to expose the roots to the air and kill the grass.

The prairie breaker was a heavy plow. The moldboard alone often weighed 125 pounds. The fourteen- or fifteen-foot beam and the handles made it even heavier, but this weight was needed since it kept the plow from bucking out of the furrow as it struck the fiberous root system. Two small wheels supported the plow beam in front and the depth of the cut was regulated by a lever which ran from the handles to the front of the beam. By lifting the lever the share would dig deeper into the soil; by depressing it, the plow could be raised from the ground. One of the front wheels ran in the furrow and was from two to four inches larger than the wheel which ran on the sod. This was necessary to keep the beam level. Sometimes, several curved rods replaced the moldboard. The rods reduced the friction on the plow while lifting and turning the sod.

Plowing with a prairie breaker was slow, hard work, even though it cut only two or three inches deep. Depending on the toughness of the root system, as many as

**Fore Carriage for Wisconsin Breakers.**

This Fore Carriage is all steel except the lifting lever and rear post. The upright steel standards are held in position on the beam by steel clips and are securely clamped to an $1\frac{1}{2}$-inch steel axle. By loosening the axle clamps the axle may be set to accommodate any width of cut from 18 to 24 inches.

The wheels are steel with three-inch tire, staggered spokes. Equipped with oil-tight, dust-proof boxes. With these boxes it is an easy matter to oil the axles by filling the screw-cap with axle grease and screwing same into place on the hub.

*Advertisement for the Wisconsin Breaker. (Smithsonian Institution.)*

*Midwestern farmers used a heavy breaking plow to turn the tough, prairie sod. The moldboard was made from highly polished iron or steel to prevent the soil from sticking. The prairie breaker only cut several inches deep. Nevertheless, three or four yoke of oxen were needed to pull it. (Smithsonian Institution.)*

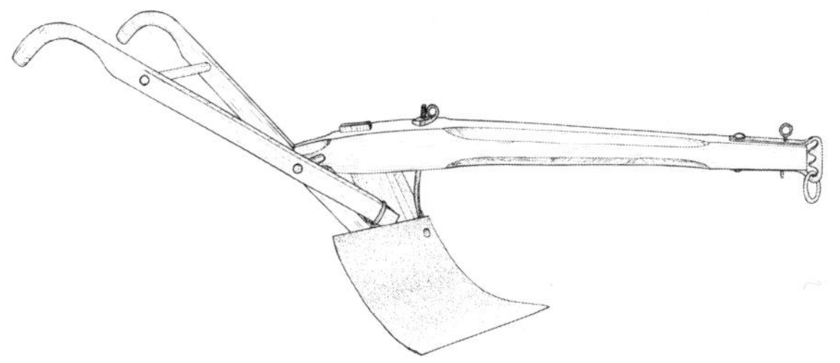

*In 1837, John Deere built a plow with a highly polished wrought-iron moldboard and a steel share. It easily cut through the prairie sod and turned a furrow. John Deere made this plow in 1838, and it is probably identical to the 1837 model. He did not make a steel moldboard until 1846. The drawing shows the entire John Deere plow. (Smithsonian Institution.)*

*From the 1840s to the 1860s, the Eagle plow was the most popular tillage implement in the United States. This plow was built in many styles, but it was characterized by a longer and more curved moldboard than that found on other plows. The Eagle plow was well suited for turning a furrow in most soils. (Smithsonian Institution.)*

three to seven yoke of oxen might be needed to pull this plow. If a farmer broke as few as eight acres during the plowing season, he was fortunate to have done that much work. A hired breaking team could usually do more, since the plowmen did not have other farm chores to take up their time. A two-man breaking team, using three yoke of good-sized oxen and a twenty-four inch plow, could turn three acres a day. The share usually had to be taken to the blacksmith once a week to be pounded thin and sharpened. In between trips to the blacksmith, the plowmen usually filed the share frequently, generally after each round on an eighty-acre field. Custom breaking teams had two shares so they would not need to stop work while a boy took the dulled share to the nearest blacksmith. Frequently, the plowmen, "cold hammered" the share when it dulled. To do this, the share was removed from the moldboard and placed on a piece of railroad iron which served as an anvil. The share was then hammered thin and sharpened with a file.

Although plowing with a prairie breaker was slow, the cast-iron plow of the 1830s and 1840s did not give the western farmer a viable alternative, because the sticky prairie soil clung to the moldboard. The cast-iron plow had two problems which made it nearly useless in prairie soils. First, it did not take a high polish. Consequently, the moldboard would not scour, that is, it tended to hold sticky soil instead of allowing it to slide or peel off as the implement was pulled through the ground. This caused constant delay while the farmer unclogged the plow by scraping it with a paddle. Secondly, the cast-iron surface contained small cavities known as blow holes which filled with the clay-like soil. This clogging caused unwanted friction which increased the draft required.

About 1833, John Lane, a Lockport, Illinois blacksmith, made the first successful effort to design a plow that would not clog after the initial breaking had been completed. Lane recognized that only steel, not cast iron, would scour suitably to permit the moldboard to turn a clean furrow. In order to produce such a plow, Lane plated a wooden moldboard and share with strips of steel cut from an old saw. Lane's innovation worked better than any plow tried in the prairie soils to that time. He did not, however, patent his idea nor produce

*By the late nineteenth century, steel breaking plows were widely available. This steel plow has a wooden beam and a disk coulter.*

*During the 1850s and 1860s, the Michigan Double-Plow became popular. The front share cut the furrow and turned several inches of soil. The larger, rear share cut deeper and turned a thoroughly pulverized furrow. (Smithsonian Institution.)*

*The sulky or riding plow became popular in the trans-Mississippi West during the 1870s. It enabled farmers to plow faster than they could with the walking plow and to ride at the same time. The sulky plow was an important tillage implement for farmers who had large acreages to prepare for seed. (Smithsonian Institution.)*

his steel plows on an extensive basis. Steel was both scarce and expensive, and it remained so until the Bessemer and open hearth processes enabled steel producers to increase production and reduce the price. Even so, other agricultural inventors began applying Lane's discovery to their own work. One of the most successful and widely recognized inventors in these efforts was John Deere.

Actually, John Deere, a Grand Detour, Illinois blacksmith, did not make steel plows on a wide basis until the mid-1850s. His early plows, the first made in 1837, consisted of a highly polished wrought-iron moldboard with a steel share. Deere cut his diamond-shaped moldboard and landside from a single piece of wrought-iron, then heated and bent it over an anvil until it took the desired shape. To this, he attached a steel share. There is no clear evidence that he fashioned his first plow from a steel saw blade. Nevertheless, this myth has been repeated so often that it is accepted as fact. More correctly, Deere's reputation for devising a steel plow came from his use of a steel share which was stronger than cast-iron and which held a sharp edge better than wrought-iron. Farmers, by calling this implement a steel plow, were probably merely using the term to distinguish it from more traditional wooden and cast-iron plows.

John Deere's highly polished, wrought-iron moldboard with steel share easily cut through the sticky prairie soils without dulling or clogging and with about half the draft which cast-iron plows required. His plow was so superior to the heavy breaking plows that it became commonly used as a breaking plow itself. Quickly, it earned the nickname "singing plow." In 1846, Deere received the first slab of cast plow steel rolled in the United States, but this metal was still too expensive to warrant extensive production of steel plows. Until the mid-1850s, most of Deere's plows consisted of wrought-iron moldboards with steel shares. Not until the 1860s, when several steel manufacturers began making consistently high quality crucible steel, did implement manufacturers begin using steel on a wide basis. Even so, steel plows cost as much as two and a half times that of cast-iron plows. Furthermore, many technical problems remained to be solved before

*Sulky plowing near Russell, Kansas, 1910. (Smithsonian Institution.)*

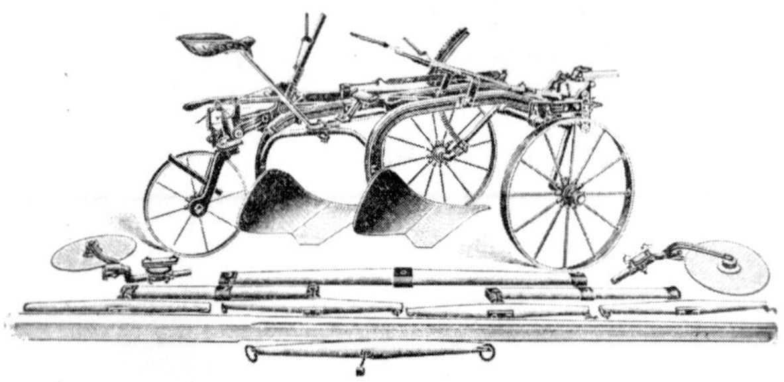


*The sulky plow became a gang plow, when a second moldboard was added. Sulky gang plows required four or five horses for draft instead of two for a single moldboard sulky.*

a completely satisfactory steel plow could be made.

Although Deere's plow was effective in the Midwest, he had manufactured and sold only a few by 1843. In that year, however, Deere added a foundry to his shop and together with Leonard Andrus, produced 400 plows. In 1847, Deere moved his shop to Moline, Illinois, and with expanded capital increased production to 700 plows. By 1857, twenty years after his first plow, John Deere was producing more than 10,000 plows annually. Over that time, he also refined the shape of the moldboard from the original curved trapazoid to one that looked more like the traditional moldboard.

Largely because of John Deere's efforts, the steel plow was in great demand by the late 1850s. Yet, while many farmers awaited the perfection of the steel plow, they still needed an efficient tool that would turn a furrow and not break or dull quickly. This meant using the best cast-iron plow possible. Fortunately, Joel Nourse, of Worcester, Massachusetts, succeeded in making a cast-iron plow which adequately broke rough ground and turned a furrow in soil with weeds and heavy stubble. This implement, called the Eagle plow, differed in two respects from all other cast-iron plows. First, Nourse lengthened the moldboard. Second, he gave the moldboard additional curvature. The result was that the Eagle plow lifted the soil and turned it over more effectively than did any other cast-iron plow. The Eagle plow cut a furrow 7 inches deep and as much as 14 inches wide depending on the model. With abilities such as these, it became a popular implement for several decades. In fact, from 1840 to 1861, Nourse sold from 25,000 to 30,000 Eagle plows annually — more than any other plow-maker in the United States.

During the 1850s and 1860s another cast-iron implement, known as the Michigan Double-Plow, came into widespread use. The Michigan Double-Plow had two shares, the smaller of which was attached to the beam ahead of the larger moldboard. The forward share pared off several inches of sod, inverted it, and deposited the slice into the previous furrow. The following share or main plow could then cut more deeply. The Michigan Double-Plow prepared a thoroughly pulverized seedbed which was excellent for all crops, but it was particularly well suited for preparing a deep, loose soil favorable for vegetable crops. The Michigan Double-Plow cut deeper than single moldboard plows, but it required greater draft power to enable both shares to cut through the soil at once. Still, while the Michigan Double-Plow was never exceptionally popular for breaking virgin sod in the Midwest, some farmers used

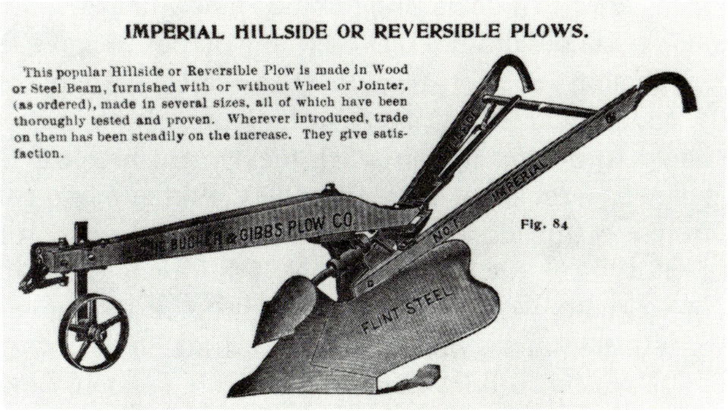


*The hillside or sidehill plow enabled farmers to plow back and forth across steeply sloping land. At the end of the furrow, the plowman unlocked the moldboard and rolled it under the beam to the other side. By so doing, the moldboard cast the furrow down hill as the farmer guided the plow back and forth across the face of the hill.*

*Farmers used the subsoil plow to loosen the "hard pan," that is, the packed soil beneath the regular cut of the moldboard plow. It was also used to deepen the furrow to permit better moisture penetration. The wheel and bracket on the beam regulated the depth of the cut.*

it for that purpose, because the second share thoroughly covered the grass roots which the first share exposed. Even so, the Michigan Double-Plow could be used only as a breaking plow in the spring when the soil was damp. At that time, three horses could plow two acres a day, but if the soil was dry the same number of horses could not budge this plow in the tough prairie sod. More appropriately, farmers found this implement useful for preparing the seedbed with a second plowing, after the initial breaking had been done and the sod effectively killed.

Even though the Eagle and Michigan Double-Plows were popular during the 1850s, prairie farmers still preferred the steel-shared or steel moldboard plows when they could get them, and steel plows were in great demand by the outbreak of the Civil War. Nevertheless, steel plows had a number of imperfections that prevented maximum effectiveness, particularly when manufacturers substituted inferior material or were haphazard with the finishing process. Certainly, steel plows turned the prairie soil better with less power than the cast-iron models, but they were expensive and either broke or wore out more quickly than cast-iron plows, if they were not tempered correctly.

A major problem with the steel plow was that the metal's quality did not permit maximum performance. Given the state of steel technology at that time, the metal could not be tempered uniformly. During the process of shaping the plow, the steel would hold its form only when it was cold. When the steel was heated for tempering, it would warp and it would continue to warp as it cooled, thereby ruining the shape of the plow. In addition, improperly tempered steel would not scour properly. In 1868, John Lane (the son of the earlier inventor with the same name) developed a process for making "soft-center" steel. Lane welded a soft bar of cast-iron between two bars of steel. He then rolled the block into a thin plate for tempering and shaping into moldboards. This process ended some of the warping problems. Moldboard plows could be fashioned now which maintained their scouring ability and which were virtually unbreakable.

Early in 1869, James Oliver, also a blacksmith, patented a process for hardening cast-iron so the moldboard would wear longer and scour better than regular cast-iron plows. The process involved passing a stream of warm water over the hot cast-iron. Oliver called the result "chilled iron," because it cooled rapidly and became exceptionally strong. Chilled iron was also cheaper than soft-center steel. In 1870, Oliver produced his first plow for commercial sale. His plows became popular nationwide, because they were light and durable and had less draft than other iron models. By 1878, Oliver was producing 60,000 plows annually, and more than 175,000 chilled iron plows were in use.

By the last quarter of the nineteenth century, the walking moldboard or swing plow operated efficiently behind a draft horse. A variety of moldboards were on the market which scoured and turned a furrow under many soil conditions. Plows with iron beams gave the implement superior strength when used in stubbly soil. Wooden beams, however, were lighter and were less likely to be sprung, although they might break if the share hit a rock that did not give way. Furthermore, cast-iron was abandoned in favor of chilled iron and soft-center steel for plow-making. Additional plow improvements would come only with the introduction of gasoline tractors which increased draft power.

In the meantime, farmers moved beyond the midwestern prairies and onto the Great Plains. Agriculture now became more extensive than ever before. Bonanza farmers also began breaking vast tracts of land along the Red River Valley in North Dakota as well as on the large farms along the Pacific Coast. The single moldboard walking plow was too slow to do all the work that was necessary. Consequently, these farmers began using sulky and gang plows to speed their work.

The first successful sulky or riding plow appeared about 1864, and it became popular among the grain producing farmers during the 1870s. This one bottom plow enabled farmers to take advantage of the extra draft power that they were losing when their horses pulled the walking plow. It also permitted them to cover more ground than ever before and to ride at the same time so that their work became faster and easier. The first sulkies consisted of a moldboard mounted on a two-wheeled frame with a seat. In the mid-1870s, however, the plow's landside was eliminated and the left wheel set at an angle in the furrow to support the plow. This innovation lightened the plow and decreased the draft required to pull it. A lever or foot peddle enabled the operator to regulate the furrow's depth without stopping the horses and getting off the plow. By tripping another lever or by stepping on the break peddle, depending on the model, the operator raised the plow from the soil. Several years later, in 1884, the three-wheeled model appeared on the market, and it found widespread acceptance among the farmers in the Upper Mississippi Valley. Occassionally, a second moldboard was added to the sulky, but this substantially increased the draft and these plows were commonly called "horse killers." A sulky plow with a single moldboard was generally cheaper and more efficient. Nevertheless, during the late1870s, and early 1880s, bonanza farmers in California and the Red River Valley of the North readily adopted sulky gang plows for their wheat lands. Four or five horses were used to pull the two bottom gang and eight horses to pull the four bottom implement. With the two bottom sulky plow, the operator could turn from five to seven acres per day.

In addition to wooden, iron, or steel moldboard plows, whether walking, sulky, or gang, farmers also used a variety of specialty tillage implements such as the hillside, subsoil, ditching, paring, and disk plows. The hillside or sidehill plow, for example, was designed to allow a farmer to plow back and forth across steeply sloping ground and cast the furrow downward. In order to accomplish this task the moldboard pivoted on its axis from one side of the beam to the other. When unlocked, it could be rolled under the beam and up on the opposite side at each end of the furrow to permit plowing back across the face of the hill. This plowing method helped prevent soil erosion since the furrows acted as miniature terraces to catch precipitation. The sidehill plow was also useful for turning furrows away from stone walls or fences.

The subsoil plow was another specialty implement. This plow had neither a landside nor a moldboard. Instead, it consisted of a narrow blade or tongue which could be set to cut at various depths. Farmers used the subsoil plow for deepening the furrow to permit greater

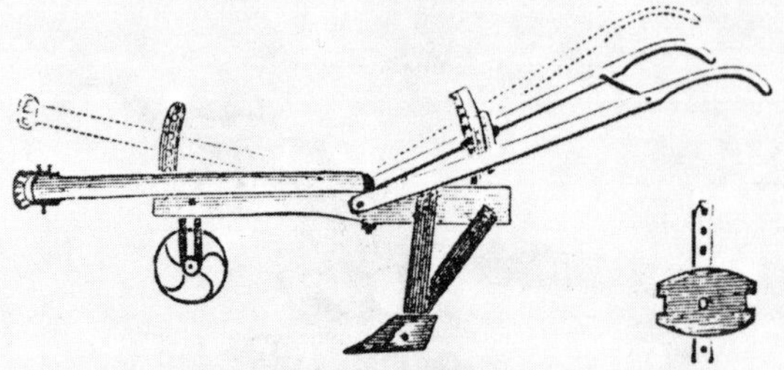

Fig. 142.—*Adjustable Ditching Plow.*

*The ditching plow helped farmers prepare a trench for drain tile. The handles adjusted to permit plowing trenches as deep as four feet.*

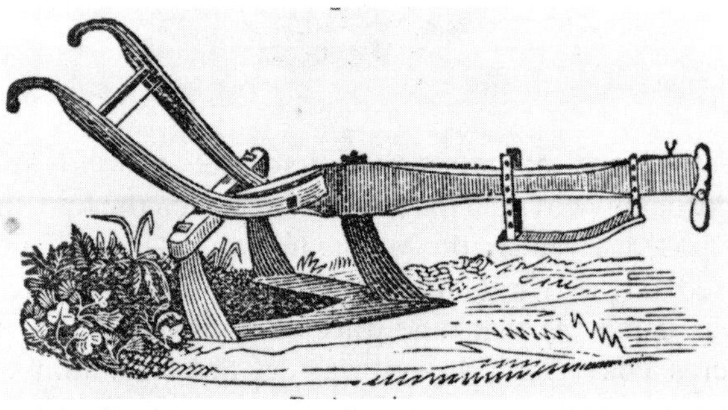

*The paring plow cut weed roots beneath the surface without turning a furrow. Occasionally, the paring plow was used to cultivate between corn rows.*

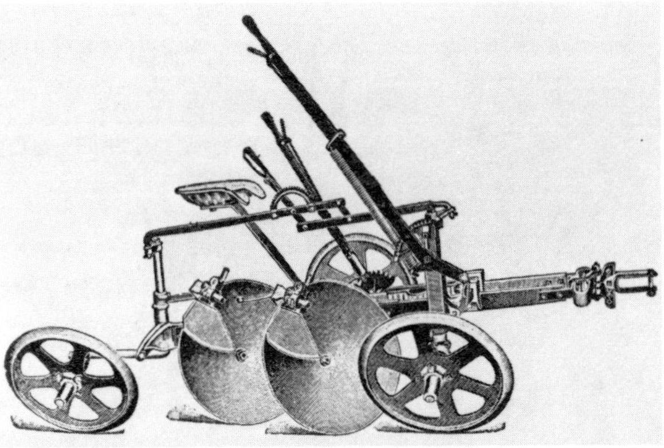

*During the late nineteenth century, Western farmers began using disk plows. The disk plow easily cut through stubble and hard or sticky soils. The disk plow, however, did not become popular until the gasoline tractor provided adequate draft power.*

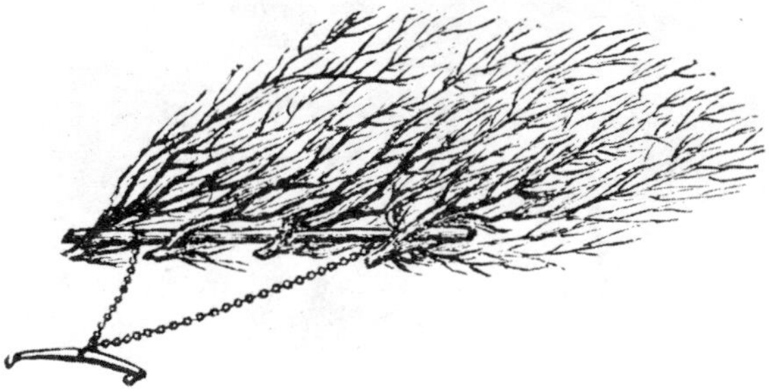

*Harrows are used to smooth the soil after the plowing, and the simplest form of harrow is a tree limb pulled behind the plow horse. Some early American farmers improved the brush harrow by attaching branches to a horizontal bar. (Smithsonian Institution.)*

moisture and root penetration. The narrow plow blade followed the furrow made by the moldboard plow. It cut from 10 to 16 inches deep without bringing the soil to the surface. Not all farmers used it during every plow season, but many midwestern farmers found it useful when the subsoil became packed or when they desired deeper plowing than usual.

When farmers had to contend with poorly drained land, they used a ditching plow to help prepare the trench for the laying of drain tiles. The ditching plow, like the subsoil plow, did not have a moldboard. Instead, it had a sharp, spiked point designed to cut a narrow furrow. This plow, pulled by two horses, cut a trench from five to eight inches deep. Workers then used narrow shovels to clean out the furrow, and the plow was drawn down the trench again, each horse walking on one side of the furrow to avoid cave-ins. Frequently, the ditching plow had adjustable handles so that it could be used at the bottom of a ditch as much as four feet deep; part of the beam was also moveable so that it swung upward while the plow was in the bottom of a deep trench. When the trench was completed and the drain tile laid, a moldboard plow was run along the excavated earth in order to turn it back into the ditch.

The paring plow was also a specialty implement. It had a flat, triangular-shaped blade which cut a three-foot swath. The blade ran a few inches beneath the surface and shaved off weeds. An adjustable gauge, attached to the beam, regulated the plow's depth. This gauge or shield rested on the ground; if lowered the plow would cut shallower; if raised it would cut deeper. The paring plow was sometimes used to cut grass and weeds between poorly cultivated rows of corn.

During the late nineteenth century some farmers, who tilled sticky or hard soils, began using a disk plow. In those soils a moldboard tended to clog or jump out of the ground. Disk plows sliced through hard and sticky soils easily and efficiently and thereby did a better tillage job than most plows. Although the the disk plow had less draft on hard ground than the moldboard, the draft proved to be heavier in proportion to the amount of work completed. The disk plow did not become popular until the 1920s, when gasoline tractors provided the adequate draft power needed for it to operate efficiently in all soils.

## HARROWS

The harrow, next to the plow, is one of the oldest and most useful implements for seedbed preparation. Plowmen considered it a necessary tool for rough, cloddy soil which the moldboard left unpulverized. Consequently, the harrow was needed to smooth newly tilled fields prior to seed planting. Harrowing encouraged an even crop stand and provided easier footing during harvest time. Harrows were also used to kill weeds and to cover seed which had been sown. The oldest and most primitive harrow was nothing more than a tree limb or clump of small trees, such as cedars, hitched behind the plow horse. Although the brush harrow did not smooth the field in a uniform manner, it did break down some of the rough spots and cover newly sown seed. An improved version of the brush harrow could be made by drilling holes in a timber and by inserting branches ten or twelve feet long in the holes. Chains could then be tied to the branches for weight, thereby, improving its smoothing ability.

By the 1790s, the American farmer was using two basic harrow styles — the square and the triangular or ''A'' frame. The square harrow was usually used on old fields that were free from obstructions such as tree stumps, roots, or rocks. The triangular harrow, on the other hand, was commonly used on newly plowed lands which had such obstructions. The triangular harrow was stronger and less likely to break, and it did not collect as much stubble as the square harrow. Both

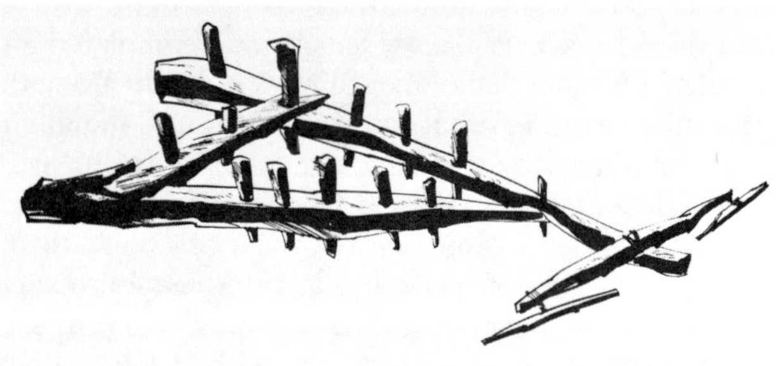

*Early American farmers made a triangular or "A frame" harrow by inserting pegs in holes bored in a timber. The triangular harrow was stronger than the square harrow, and it was used primarily on newly plowed land. (Smithsonian Institution.)*

*During the late 1860s, the Nishwitz harrow became popular. The steel disks easily smoothed roughly plowed ground. The Nishwitz harrow adjusted to various widths by moving the center crossbar.*

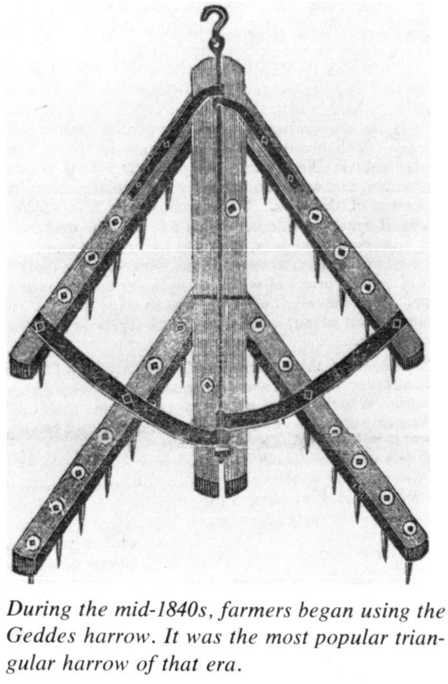

*During the mid-1840s, farmers began using the Geddes harrow. It was the most popular triangular harrow of that era.*

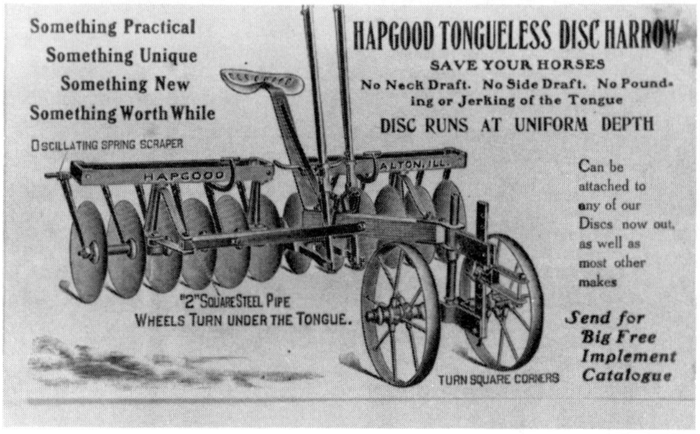

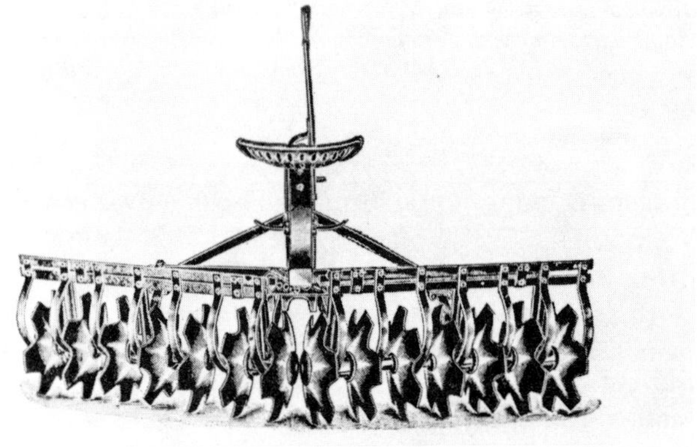

FIG. 56—A SINGLE-LEVER CUTAWAY DISK HARROW

*The disk harrow increased in popularity during the 1890s. Sometimes the disk harrow was used as a plow, but these implements did not provide maximum tillage efficiency until gasoline tractors furnished adequate draft power.*

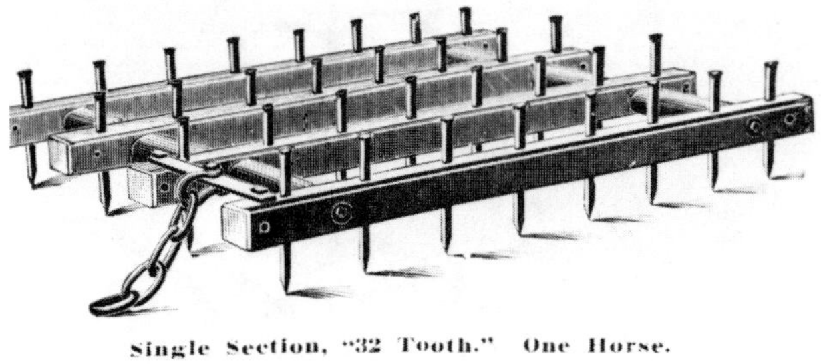


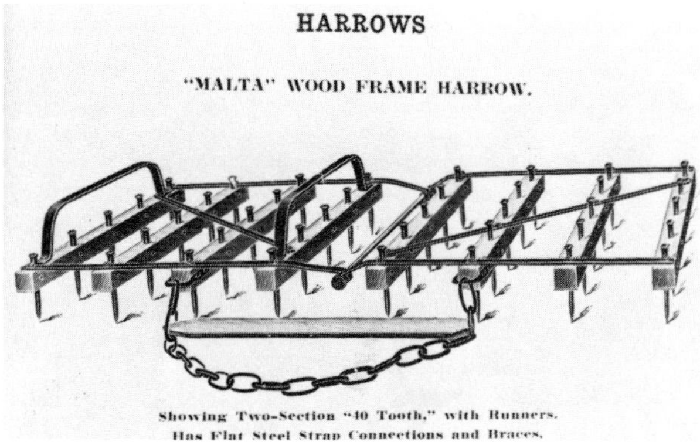


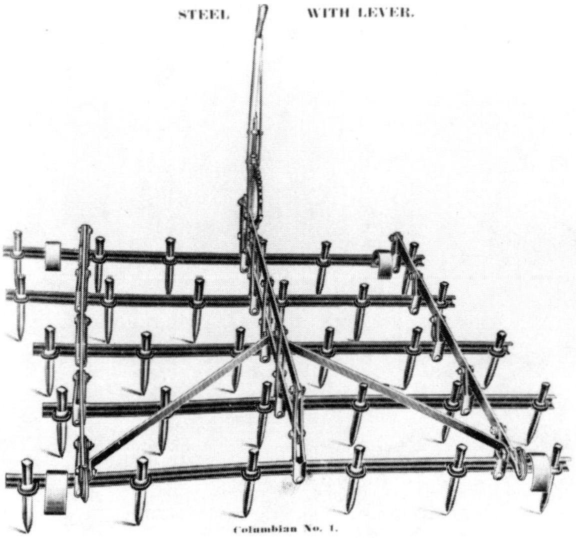


*Above: After the mid-nineteenth century, farmers used spike-tooth harrows with either wood or iron frames. The teeth were made from iron or steel. Some harrows had adjustable levers which regulated the angle and depth of the teeth.*

*Right: Some harrows, such as the Acme, had jagged, curved iron strips for teeth. No matter what type of teeth the harrow had, however, its purpose was to smooth roughly plowed ground.*

harrows had wooden frames with either wood or iron teeth.

In the 1840s, farmers began using a hinged harrow. This design worked better on rough land. It was less likely to break, since the hinge gave this implement some flexibility. The most popular triangular, hinged harrow appeared about 1845. This Geddes harrow had a light draft and it easily could be lifted to clear away clogged stubble. The hinged, square harrow was still most commonly used on well cleared land. When the square harrow was joined with another to smooth a six-foot swath, a farmer could harrow as much as ten acres a day. Still, there were problems. Iron teeth frequently broke when they struck rocks or other solid objects. Consequently, farmers had to replace harrow teeth frequently. During the 1860s, however, steel teeth were substituted for iron; manufactures also began making the frames from iron instead of wood. These two changes substantially increased the harrow's strength and performance. At that same time, harrows were designed with levers which allowed the farmer to change the pitch or angle of the teeth, depending on the nature of the ground being worked and the degree of smoothness desired.

Soon after the Civil War the Nishwitz disk harrow achieved a degree of popularity. This harrow consisted of a wooden frame held together by a cross bar. The bar adjusted to expand or contract the harrow to the desired width. The cast steel disks were about one foot in diameter and cut several inches deep. Disk harrows such as the Nishwitz and other models which came later did not pack the soil as did some harrows with iron or steel teeth. For most soils, the disk harrow was the most economical implement, since it sometimes could be used as a substitute for the plow, particularly if the soil was loose or where only shallow tillage was required. Although harrows with 14- and 16-inch disks increased in popularity during the 1890s, they were not entirely satisfactory until tractors provided adequate draft power to pull them through heavy soils. In the meantime, farmers preferred the spring-tooth harrow.

The first patent for a spring-tooth harrow was made in 1869. The spring-tooth harrow was well suited for rough, rocky ground, since the steel teeth flexed and did

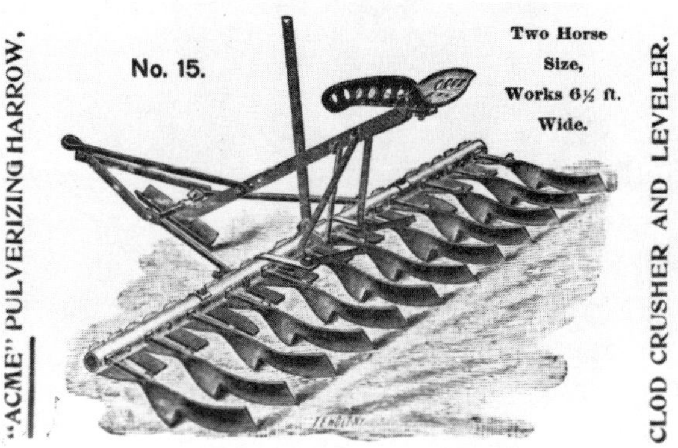

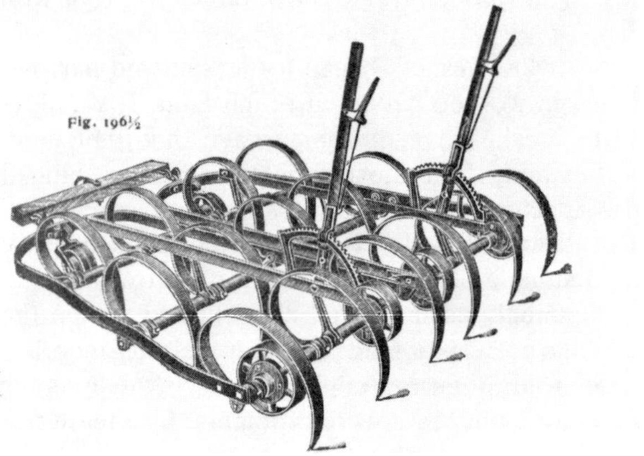


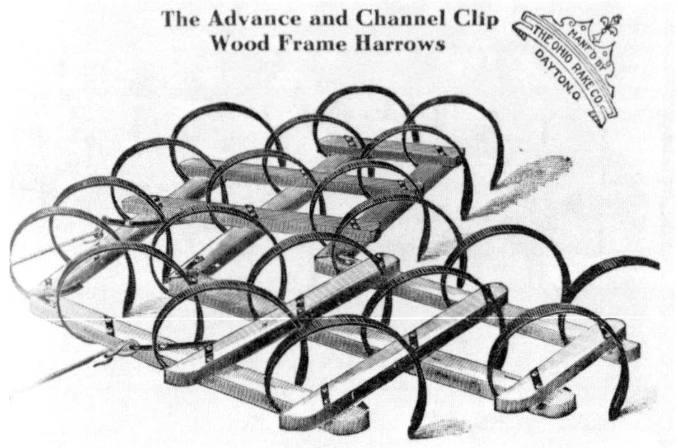


*By the mid-1880s, many farmers preferred the spring-tooth harrow. The teeth of this harrow were made from spring steel and each flexed over obstructions without breaking. Spring-tooth harrows were made with either wood or iron frames.*

*Rollers and clod crushers were made with steel or corrugated bars. These implements were used to crush large chunks of soil, left by the plow, before planting the crop.*

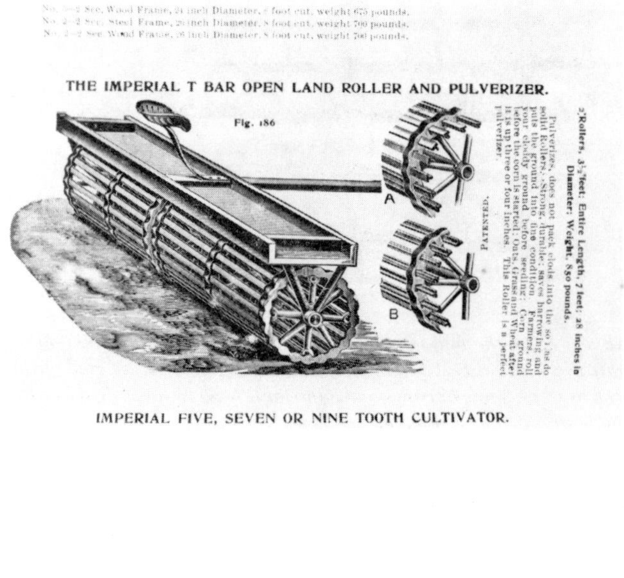

not break as did the spike-tooth models. A lever adjusted the angle and depth of the teeth. The spring-tooth harrow was also well adapted for covering seed after sowing and it required far less draft than the disk harrow. By the mid-1880s, the spring-tooth harrow was in popular demand, and with the larger models a farmer could cover from twenty to twenty-five acres a day. At that same time, bonanza farmers along the Red River in North Dakota began using spike-tooth harrows that were twenty feet wide. These harrows had four sections to permit flexibility on uneven land. An operator with a four-horse hitch could cover forty acres per day with those implements.

## ROLLERS AND CLOD-CRUSHERS

Once the plowing had been completed, some farmers occasionally used a roller or a clod-crusher to break down chunks of soil which the plow had not pulverized as it turned the furrow. The first rollers or clod-crushers were simply logs with pins driven into each end to which straps, leading to a yoke of oxen, were attached. Log rollers were difficult to manipulate and to turn at the end of a field, and they did not function properly on uneven ground. These problems were solved by the invention of iron rollers made in two or three sections. Each section raised or lowered independently as the implement passed over uneven ground. The most effective clod-crushers consisted of cast-iron disks which were loosely fitted over an axle so that each disk revolved separately.

Clod-crushers and rollers tended to pack the soil, but rollers did so more evenly. Consequently, farmers sometimes used this implement to pack the soil after the seed had been planted to enhance germination. More often than not, however, packed soil was a hindrance because it enabled moisture to escape through tiny, capillary-like holes. In order to prevent that moisture loss, farmers harrowed their fields after they used a roller or clod-crusher. The harrow broke up those capillary passages and caused the moisture to remain in the soil for a greater length of time. Rollers and clod-crushers were primarily used on heavy soils that were not well drained. In wet soils, the plow invariably turned over large chunks of earth that had to be broken down before seed could be planted. Rollers were particularly well suited for preparing a smooth, level seedbed which was beneficial to farm hands who came later with a grain cradle, grass scythe, hay rake or mower. After a farmer installed drain tiles, however, the soil dried better and the plow pulverized it more efficiently. Consequently, rollers and clod-crushers were seldom used.

With these implements, then, the farmer stirred, turned, and smoothed the soil for the preparation of the seedbed. Few other major technical changes occurred before the adoption of steam and gasoline tractors which required plows and harrows to be designed for rapid travel through the soil. Until then, once the plowing and harrowing was completed, the field was ready for planting. The increase in technical knowledge, which enabled inventors to improve the plowman's tools, had even greater impact upon seeding and planting implements. Indeed, technological advance revolutionized seedtime for the American farmer.

*Prior to the perfection of the grain drill, farmers either seeded their crop by hand or they used a broadcast seeder. The handcrank broadcast seeder consisted of a bag with a star-shaped disk at the bottom. As the crank turned, seed fell onto the spinning disk and scattered across the furrows. (Smithsonian Institution.)*

# CHAPTER III

# SEED TIME

From antiquity until the mid-nineteenth century, farmers planted seed by hand. Some broadcast their seed, that is, they scattered it in the air as they walked across their fields. If the seed was sown broadcast, however, the field had to be harrowed in order to cover as much of the grain as possible, and, thereby, protect it from the birds, wind and weather. Still, an even stand of grain was difficult to obtain by broadcast sowing, because the amount of seed which fell to the ground at each cast depended upon the sower's skill and the force of the breeze or wind.

## GRAIN DRILLS

About 2,000 B.C., Babylonian farmers began experimenting with a seed drill which required two men to operate. As one farmer guided the drill across the field, a second man dropped seed into a pipe which deposited it in the trench behind the furrow opener. The Assyrians experimented with a similar seed drill during the seventh century B.C., but this implement remained virtually unchanged until the late sixteenth century A.D. At that time, Italian inventors added a revolving mechanism which automatically dropped the seed into the furrows.

Few other effective improvements were made in the seed drill until the early eighteenth century, when Jethro Tull, an English inventor, began experimenting with it. In 1733, Tull published *Horse-Hoeing Husbandry* in which he described and illustrated a seed drill. This one-horse drill seeded three rows of wheat or turnips at once. Hoe-shaped coulters or points opened the furrow. The seed dropped into the soil through a tube which passed through the coulters. These rows, spaced seven inches apart, could then be weeded with a

*The wheelbarrow seeder dropped the grain from the seed box as the wheel turned. Then, the seed had to be covered with a harrow.*

*Seymour's Broadcast Sower was popular during the 1840s. It would scatter all sizes of seeds from peas to clover, and it would spread lime as well. The lever above the right wheel adjusted the seed opening at the bottom of the box.*

horse-drawn cultivator. Other grain drill innovations followed Tull's, but few were more than marginally satisfactory.

American farmers began using the seed drill on a very limited basis about the time of the Revolution. But as was true with the plow, these drills were either imported from England or they were made locally. Newly cleared fields with an abundance of stumps and rocks did not lend themselves to the use of this implement. Furthermore, most farmers seeded winter wheat in fields where corn had recently been harvested and the stalks plowed under, and the corn stalks and weed stubble clogged in the drill tubes. These early drills also failed to plant uniformly on roughly plowed ground. In addition, seed drills were too expensive for most farmers, and for about the first decade, the soil was usually so fertile that farmers could reap bountiful harvests simply by broadcasting seed. Finally, those farmers who were familiar with grain drills did not find them practical because they could sow as much grain by hand in a day as they could reap by hand at harvest time.

Even so, the first American patent for a grain drill was made in 1799 by Eliakim Spooner of Vermont. The patent drawings for this tool have been lost, but no matter what the implement looked like, it evidently did not function very well because it was not reproduced on a wide basis — if at all. Other patents for seeders of various kinds followed during the next forty years, but little technical advance was made, and few farmers had any direct knowledge of grain drills or their use.

In the meantime, most farmers either sowed by hand or used broadcast seeders. These seeders were of three general types — handcrank, fiddlebow or wheelbarrow. The handcrank and fiddlebow seeders consisted of a bag with a star-shaped or vaned spreading device and a hand crank or a bow. The seed bag was carried with the aid of a shoulder strap, and as the farmer walked across the field, he turned the crank or worked the bow back and forth. By so doing, seed fell from the sack onto the spinning disk, and it was, in turn, cast out across the ground in a radius of several feet. These seeders were satisfactory only for small fields. The evenness or uniformity of the crop stand depended upon a consistent turn of the crank or pull of the bow and the operator's walking speed, but frequently, the seed was sown too thinly or too thickly. The wheelbarrow seeder had a long seed box which extended several feet laterally beyond the frame. The wheel meshed with a set of gears which powered a revolving agitator inside the seed box. The agitation forced the seed through the holes in the bottom of the box and onto the ground. Then, it had to be covered by a harrow.

Seymour's Broadcast Sowing Machine, patented in the early 1840s, was more effective than either the handcrank, bow or wheelbarrow seeders. The Seymour broadcast seeder consisted of a long box with an adjustable opening which extended the length of the bottom. As the seed fell from the box through the opening, a vibrating rod, powered by a sprocket chain from one of the wheels, scattered the seed. Since the seed fell only a short distance to the ground, it did not scatter much in the wind, and since it dropped seed more consistently than the handcrank models, a more uniform crop stand resulted. This one-horse, box seeder gained limited popularity in the wheat producing regions of New York and the Middle Atlantic states, but it never won wide acceptance among wheat farmers in the Midwest.

In 1841, however, Moses and Samuel Pennock of Chester County, Pennsylvania, made a major improvement in grain drill design. The Pennock drill was fitted with seven hoes and seed tubes spaced nine inches apart. Seeds in the hopper fell into a cylinder where a flanged shaft caught them as it revolved. As the

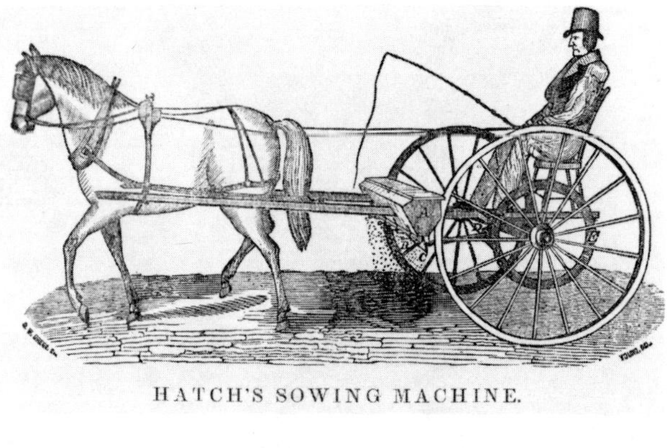

HATCH'S SOWING MACHINE.

Palmer's Wheat Drill was introduced to the grain farmers of New York about 1848. Notice how the drill tubes were designed to spring over obstructions in the field.

*The "force-feed" grain drill depended upon this mechanism. Seed fell from the grain box into the grooves of the cylinder. As the cylinder turned, it emptied the seed down the drill tubes. Force-feed grain drills planted the crop more uniformly than sowing machines or drills which did not have this device.*

*Below: The endgate seeder attached to the rear of a wagon. A sprocket chain linked the seeder to the left rear wheel. The operator filled the seeder's bin from the basket in the wagon bed. (National Archives.)*

shaft rotated, the seeds fell down the tubes and into the furrow. A series of levers enabled the farmer to raise or lower each drill tube separately. The Pennock drill reportedly worked well on rough or smooth ground. With it, a man and two horses could sow ten to twelve acres a day with a variety of seeds such as wheat, rye, corn, oats, peas, beans, rutabagas, or turnips. It could also be regulated to drop any amount of seeds per acre that was desired. The Pennock drill was manufactured and used on a small scale in Chester County, Pennsylvania, and New Castle County, Delaware, during the planting season of 1841. By 1850, this drill was in wide use in the wheat lands from Virginia to New York and west into Ohio, and it remained one of the most popular grain drills up to the Civil War.

A major improvement in planting devices came in 1851 with the patent of the first "force-feed" grain drill. Drills of this design had a notched disk or spindle which collected and metered the same number of grains for each seed release. The amount of seed dropped into the furrow depended on the size of the notched disks or the speed at which the disks rotated. The disks revolved in a cylinder which had an opening on top. As the seed fell from the hopper into the cylinder, the disk turned, caught the seed, and passed it into the grain tubes. Nearly twenty years later, in 1868, C. O. Gardiner of

*This grain drill patent model has furrow openers called "hoes." The seed fell from the grain box through tubes into the hoes. The levers enabled the farmer to set the planting depth and position of the hoes. (Smithsonian Institution.)*

*The P.P. & Mast Company of Springfield, Ohio, manufactured this eight-hoe Buckeye Grain Drill. The gearing for the force-feed mechanism is located on the center of the axle.*

Springfield, Ohio, patented a force-feed grain drill which was produced by the Thomas & Mast Company. This popular implement became known as the "Buckeye." It had telescoping seed tubes which could be raised or lowered to facilitate easier traveling on country roads.

In 1857, a grain drill was patented with a furrow opener consisting of two disks which were set at an angle to open the trench in front of the seed tubes. Chains or "press" wheels, placed behind the tubes, covered the seed. Although the disk drill did not achieve immediate popularity, it was well suited for hard, crusted, or stubbly soil. During the 1880s, Great Plains and Red River Valley farmers in the North began using the single and double disk drills.

By the mid-1860s, the drill was in common use among mid-Atlantic, Pennsylvania, and Ohio farmers. A decade later about half of the winter wheat was drilled in the Mississippi Valley, and spring wheat farmers were beginning to use it on an increasing basis. By the early 1880s, 57 percent or 14,000,000 acres of winter wheat lands in 24 states were seeded with the drill — an estimated 10 percent increase since 1874.

The grain drill became even more efficient and widely used about 1885, when a "shoe" type furrow opener was added to the seed tubes on some models. Flat or coiled springs pressed the shoes into the ground. These springs gave the grain tubes flexibility and made them less likely than the hoe drill to clog in stubbly soil, but it could not penetrate the soil as well.

Farmers along the Pacific coast at this time, however, still preferred the broadcast seeder which was designed to fit on the end of a wagon. This endgate seeder came into use about 1859 along the Pacific coast. It was used on a more limited basis in the spring wheat region of the Upper Mississippi Valley during the 1870s. The endgate seeder consisted of a large seed box with several scattering fans. A sprocket was attached to one of the wagon's rear wheels. A chain connected the sprocket to the seeder, and as the wagon moved forward, the fan turned and cast the seed. Although farmers could scatter seed about 60 feet and sow twenty to thirty acres a day, it had the same limitations of the smaller hand cranked seeders. The slightest breeze scattered the seed unevenly and a poor crop stand resulted.

During the 1860s, this problem was somewhat solved when implement manufacturers attached a force-feeding device to the box seeder. The force-feed mechanism consisted of a fluted cylinder attached to a shaft. Seeds fell from the hopper into the grooves. The cylinder revolved and emptied the seed into the tubes and onto the ground directly below the seeder. The combination seeder-cultivator also achieved popularity during the latter nineteenth century. This box seeder consisted of an agitator powered by the ground wheels. A trailing row of cultivator teeth covered the seed. The one-horse seeders covered about four acres a day, while the two-horse models sowed about six or seven acres. On the bonanza farms of the Red River Valley, eight-foot, seeder-cultivators covered as much as twenty acres per day, while the fourteen-foot, six-horse models covered up to forty acres, but these estimates may be rather high.

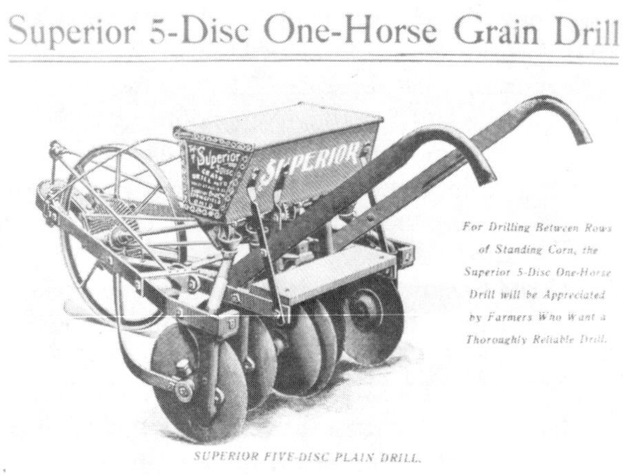


*One-horse grain drills were economical for seeding small fields. Each disk opened the furrow for the seed tube. The gearing on the wheel regulated the amount of seed which fell through the drill tubes.*

*Large, double disk grain drills, such as this twenty-row Superior, were well suited for seeding large fields in the Great Plains and Far West. The spring attached to each furrow opener enabled the grain tubes to lift over obstructions.*

Certainly the grain drill and the seeder-cultivator combination eliminated a second trip across the field to harrow under the seed. Still, the seedbed had to be smoothly prepared for these implements to work effectively. Even then, many of the grain drills did not measure seed accurately. Frequently too much or too little seed passed through the seed tubes or they clogged easily. The force-feed mechanism and the disk furrow opener, however, greatly alleviated these problems. Drills, however, generally produced greater yields and used about one-quarter a bushel less seed per acre than

*Above: During the 1860s, implement manufacturers began adding the force-feed mechanism to the box seeder or sower. The force-feed mechanism deposited a uniform amount of seed or fertilizer through the tubes. Here, a force-feed sower evenly spreads fertilizer across the field. (Smithsonian Institution.)*

*Left: During the late nineteenth century, many broadcast seeders had a trailing row of cultivator teeth which stirred the soil and covered the seed.*

did broadcast seeders. Grain drills planted seed evenly and at a uniform depth, and the furrows helped protect the new plants.

Certainly, drills saved a great amount of labor. In 1830, for example, one man could sow an acre of wheat in an estimated time of one hour and twenty minutes, or roughly seven to fourteen acres in a ten-hour day using one to two and a half bushels of seed per acre. Pennock's drill, however, might plant as much as fifteen acres per day. By the 1890s, fourteen- to sixteen-foot drills pulled by four horses easily planted fifteen to twenty acres a day. These figures are only rough estimates, because drilling rates varied with the soil conditions, the number of horses used, and the speed at which the farmer wanted to work. A general estimate for drilling wheat, however, is that a two-horse drill would plant nine and a half acres a day, a three-horse drill eleven and a half acres, and a four-horse drill about sixteen acres per day. In 1890, most wheat farmers were using grain drills, if their fields were fairly level and clear from obstructions. At that time, wheat farmers believed grain drills would pay for themselves, in terms of increased yields and labor saved, in one year.

*The Randall and Jones Double Hand Planter enabled a farmer to seed his crop in evenly spaced rows for easier cultivation. (Smithsonian Institution.)*

## CORN PLANTERS

Until about 1850, the American farmer planted corn just as the Indians had taught the first colonists to do nearly two and a half centuries before. Planting was done by hand with a hoe or a pointed dibble stick. With the hoe, the farmer chopped a hole and dropped a few seeds into the ground. He then formed a hill of earth over the seeds to help support the young stalks. If he used a dibble stick, the farmer simply made a hole, dropped in the seeds, and covered them with the heel of his boot. The corn farmer could plant about one-half to one acre a day with the hoe, or approximately ten acres during the planting season. Since the planting season varied from three to five weeks and since the corn had to be cultivated by hand, time and climate, in the absence of mechanization, severely limited corn acreage nationwide.

During the 1850s, however, the hand cornplanter appeared on the market. This planter consisted of two wooden slats with handles and a seed canister attached. A slide joined the slats in the middle. Two sharp pieces of metal, which opened or closed as the slats were pushed together or pulled apart, joined the bottom end. The slide passed under the seed canister and a small hole in it filled with seeds as the slats were pulled apart. The point was then thrust into the ground and the handles closed. This action caused the seeds in the slide to drop to the point of the planter and fall into the

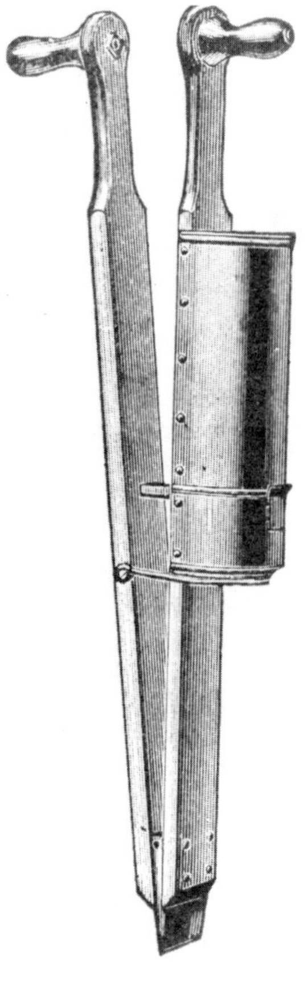

*The American farmer began using the hand corn planter or "jobber" during the 1850s. It was used primarily to plant gardens and to reseed small areas in the field where poor germination occurred.*

*Above: During the 1860s, the check-row planter became popular among corn farmers. Two runners or "shoes" opened the furrow. As the farmer drove the implement at right angles to a previously marked field, a boy tripped the handle and the seed fell at each intersection. Check-row planting enabled the farmer to cultivate both directions between evenly spaced rows. (Smithsonian Institution.)*

*This cross section of a double-row corn jobber shows on the right the seed canister and several kernels ready to fall from the point. (Smithsonian Institution.)*

ground. The hole was then covered with the farmer's boot. Many styles of this corn planter were manufactured but the tool never gained widespread acceptance because farmers did not think it saved enough time. A two row planter such as the Randall and Jones Double Hand Planter was more popular, but hand corn planters were primarily used to seed small fields or gardens and not large acreages, or for reseeding spots where poor germination occurred.

The development of relatively satisfactory seed drills, such as the Pennock drill, encouraged inventors to apply the same principles for seeding small grains to planting corn. Actually, innovation in this respect had been going on at least since 1839. In March of that year, D. S. Rockwell patented a four-wheeled corn planter which had a shovel-type furrow opener attached under a seed hopper. A slide, operated by the rear wheels tripped at the appropriate interval and dropped the seed into the furrow. The rear wheels covered the seed and packed the furrow. Although Rockwell's concept became the basic design for the corn planter, his various models failed to work properly.

A major problem which hindered mechanization of corn planting was the practice of many eastern farmers of steeping their seed corn in tar before planting, because they believed the tar would discourage birds from eating the newly planted seed. But, seed corn treated in this manner did not lend itself to machine planting, since it tended to gum-up the planting mechanism. Nor did farmers with rocky, stumpy, or rooty lands find either the hand or the horse-drawn planter useful. In the Midwest where these problems were less severe, farmers readily adopted the horse-drawn planters of various styles and makes. Even so, some farmers in the Midwest refused to use the corn planter and continued to plant seed with the hoe. They argued corn planted in the furrow was harder to cultivate than that planted in hills, since they could not plow both ways between rows.

The objections of the midwestern farmer to corn planters were largely removed in 1853 when George Brown of Tylersville, Illinois, marketed a two-row, horse-drawn implement. This dropped seed into the furrow by a mechanism geared to the ground wheels. Five years later, in 1858, Brown added a "shoe" furrow opener for cutting through stubble-filled or crusty soil. A seed tube extended from the hopper down the length of the furrow opener. As the wheels turned, the seed fell from the heel of the shoe into the furrow.

In 1860, Brown patented further improvements of his corn planter. His planter now called for two runners or furrow openers to support the front while two wheels upheld the frame in the rear. The shoe was given a new design so that it curved upward in the front thereby allowing it to climb above and break through hard soil. A hand operated dropping device was added so that an attendant (usually a boy riding on the seat in the front of the machine) could trip the seed release at the proper moment. A lever also allowed the driver to raise the frame and seeding devices. Horse-drawn planters of this type gained rapid acceptance during the 1860s in the Midwest. With it, a farmer could plant from twelve to twenty acres a day, or approximately as much as twenty times more than he could plant with a hoe.

About this same time, J. H. Rider of Wilton, Iowa, devised a corn planter mounted on sled runners. The driver operated the seed dropper with a foot pedal and planted two rows at one time. Without wheels, however, to trip the seeding device with the aid of a cam or sprocket, this planter could never become automatic, since the operator always had to trigger the mechanism himself. The major advantage of this planter, however, was that it could be cheaply constructed. Other innovations followed, most of which were unsuccessful, corn farmers preferred Brown's model.

Even so, these early corn planters had no marking device to allow farmers to plant in parallel rows. Straight rows, an equal distance apart, were necessary to permit cultivation with horse-drawn equipment. During the 1860s, though, farmers began using large marking sleds in addition to the shovel plow to help lay out the rows. These sleds had equally spaced teeth made from iron or wood; some had wooden runners. As the sled was drawn back and forth across the field, it left parallel rows. The farmers, then, had only to plant at right angles to those rows and drop the seed at each intersection. This left a checkerboard pattern which permitted cultivation from each direction. Some farmers also began using a marker which attached to the corn planter. These markers were simply arms which extended to the side of the planter. A disk or spike cut into the soil and left a mark to guide the operator on his return trip across the field.

In February, 1857, Martin Robbins, a Cincinnati inventor, patented the first corn planter that would drop the seed automatically in evenly spaced rows. This planter attached to a jointed rod or a chain with metal buttons, which, when pulled through the seeding mechanism, tripped the dropper. The chain was staked

*This patent model of Kuhn's check-row planter (1880) shows an open seed box on the right. The wooden handle tripped the seed release mechanism in both boxes. The center lever regulated the planting depth of the furrow openers. (Smithsonian Institution.)*

*Before the farmer seeded his crop with a check-row planter, he used a marker sled to cut parallel lines across the field. Then, he planted at right angles to these marks and dropped the seed at each intersection.*

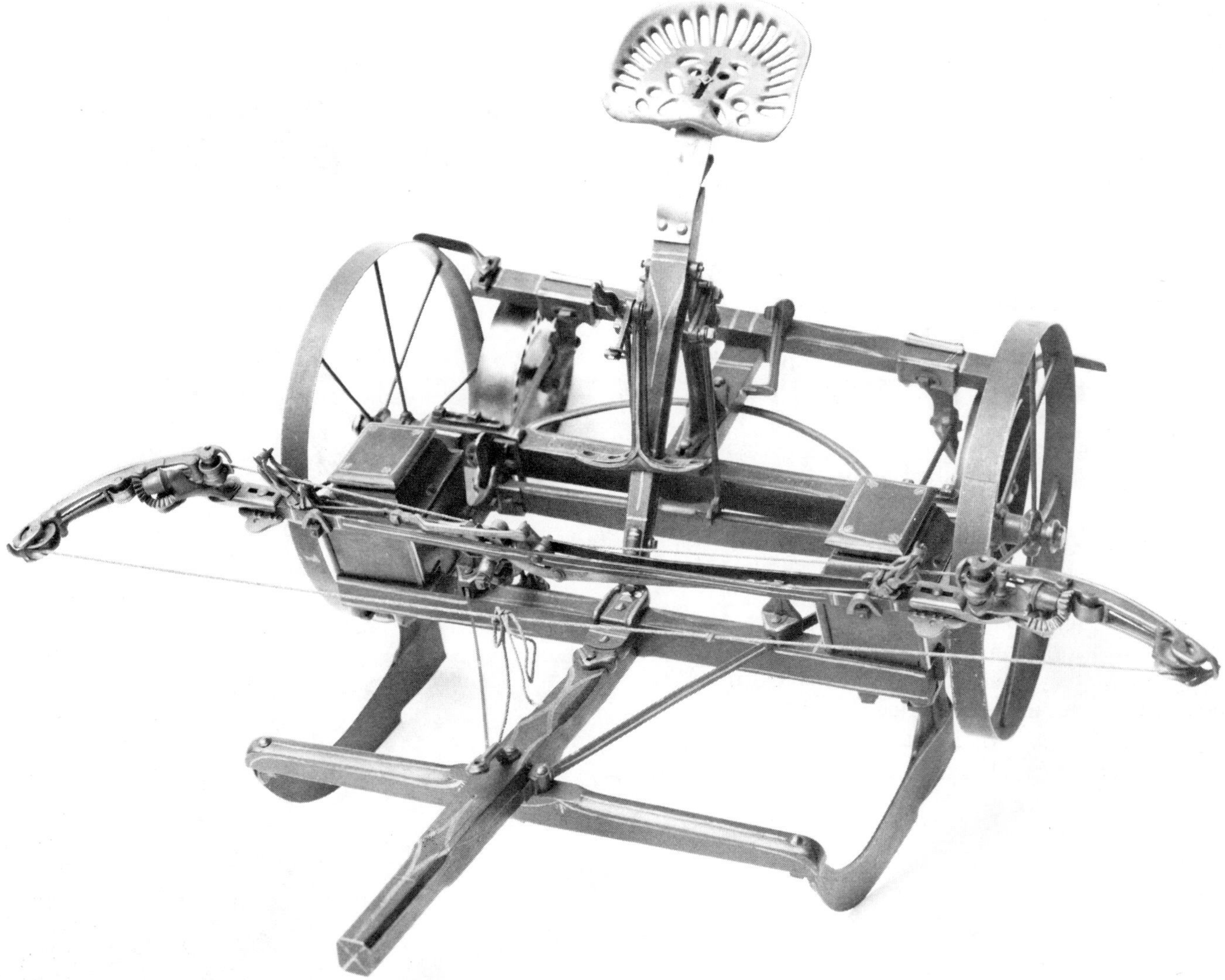

*During the mid-1860s, corn farmers began using the wire operated, check-row planter. This patent model shows the mechanisms on each side through which the wire passed. Whenever a knot passed through the pulley, it tripped the seed release gearing. Check-row planters seeded from twelve to fifteen acres per day. (Smithsonian Institution.)*

down at each side of the field, and the corn planter followed the chain as a guide. Although Robbin's planter failed to work properly, it provided the basic concept for the check-row planter which other inventors perfected.

In 1864, John Thompson and John Ramsay of Aledo, Illinois, patented a corn planter which used a knotted wire, in place of Robbins' rod or chain, to trip the seed dropper. In 1875, this patent was reissued and assigned to the Haworth brothers of Chicago for manufacture. Only one person was needed to operate this check-row corn planter. As a knot passed through the machine, it triggered the mechanism which released the seed into the tube behind the furrow opener. The check-row planter became the standard corn planting implement during the 1870s. By the turn of the twentieth century, other modifications and improvements had been made, such as allowing the wire to transfer from one side of the machine to the other as it turned around for planting back across the field. Rotary dropping devices also replaced slides which measured the seed. This too speeded the planting process.

On the western edge of the prairie and the eastern edge of the Great Plains, though, the annual rainfall was less than in the more humid regions farther east, and corn growers, planted deeper to insure proper germination. Neither single-row corn drills nor the check-row corn planter could penetrate deep enough to meet their needs. In the mid-1870s Great Plains farmers developed the lister planter which solved their problem. The lister is essentially a double moldboard plow which split the furrow and turned the slice both ways. A seed canister was attached behind the moldboard. On the walking listers, a sprocket wheel, attached behind the moldboard, rotated on the ground and tripped the seeding mechanism. Small cultivator blades or disks covered the seed. On the riding lister models, the seed mechanism was connected to a drive chain attached to a sprocket on the axle. The deep furrow which the moldboard left helped retain moisture and protect the young plants from the hot, dry winds. Lister-planted corn was also easy to cultivate because of the wide space between furrows which could be tilled with horse-drawn implements. In the corn region of the Midwest, however, plowing and planting were completed separately until the gasoline tractor provided the extra draft power needed to pull a combination implement through the heavy, sticky soils.

By the turn of the twentieth century, grain drills, seeders, and corn planters had made the planting season shorter and easier than in the days of hand power. Whether drawn by horses or by steam traction engines, these implements had become standard equipment among wheat and corn farmers. Grain drills, seeders and planters meant farmers could seed more acres of small grains and corn than ever before. With this planting technology, they would have far surpassed their ability to cultivate and harvest their crops had not technological change been just as revolutionary for these farm tasks.

*This patent model of the Hendrick's one-row corn planter (1872) shows the gearing on the front wheel which regulated the seed release mechanism under the canister. (Smithsonian Institution.)*

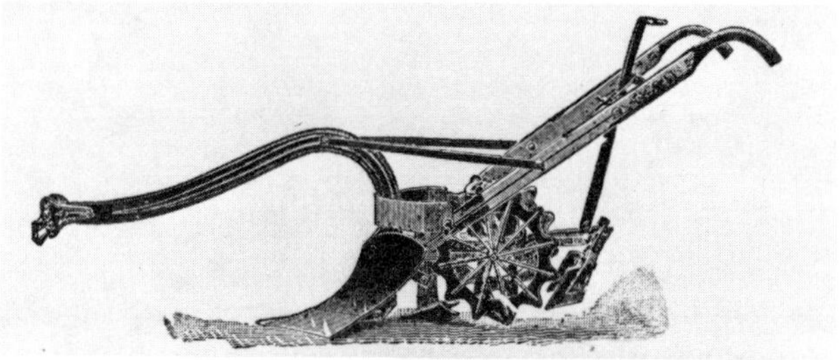

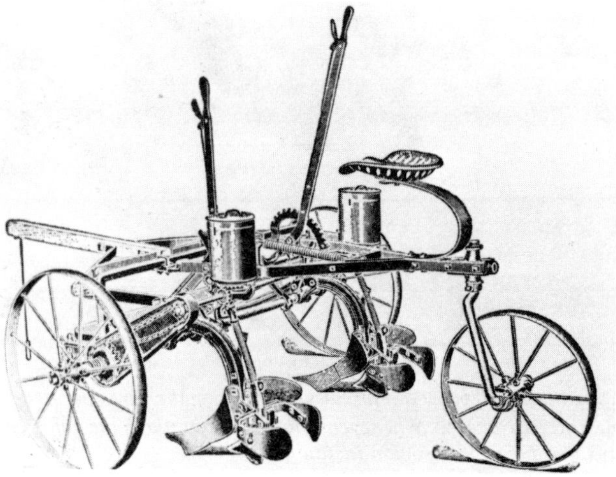

*Above: Lister corn planters, such as these single-row walking, one-row sulky and two-row sulky models, enabled farmers to plant their seed in deep furrows. The furrows protected the plants from the wind and retarded the loss of precipitation. The sprocket wheels on all three models tripped the seed release device.*

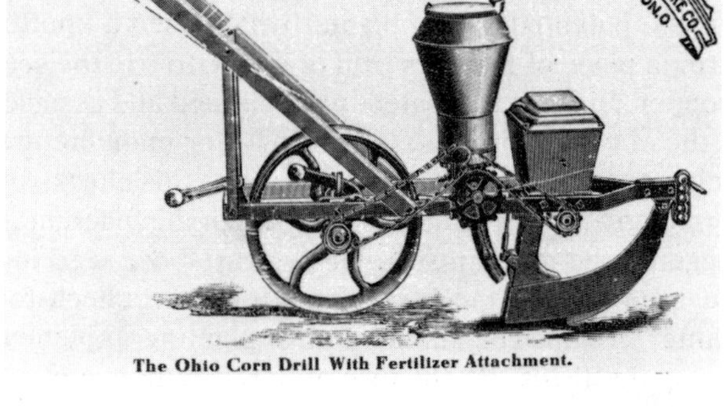


*Right: The one-row, horse-drawn planter was used to plant small fields. On this model, the chain-linked gearing regulated the planting mechanism beneath the seed box and fertilizer canister. The clevis on the front of the beam regulated the planting depth depending on how the horse was hitched to it.*

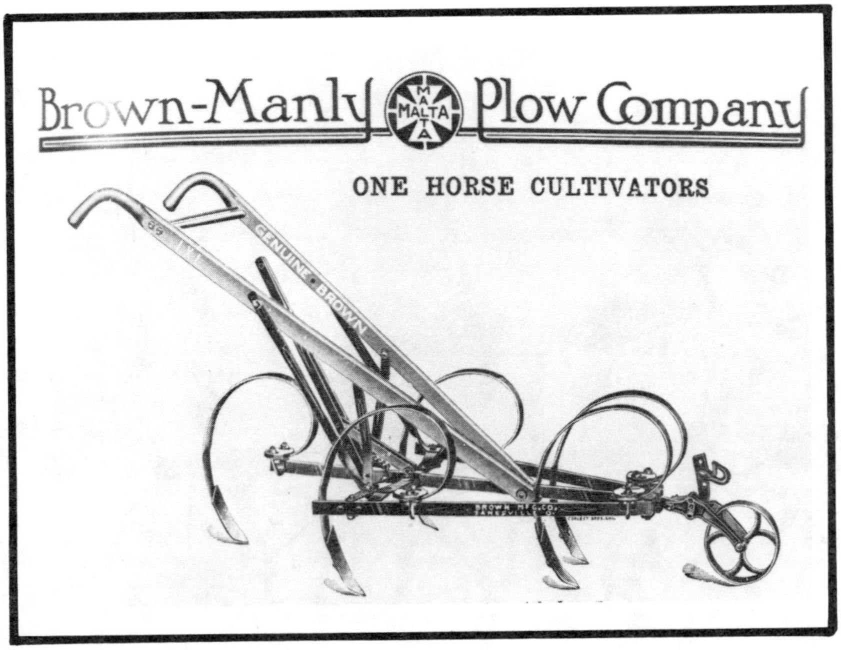


*The expandable, one-horse cultivator appeared on the market, during the 1820s. By the late nineteenth century, the walking cultivator could be purchased with spike- or spring-teeth or shovels. The spring-tooth model was popular among farmers who cultivated small fields.*

# *CHAPTER IV*

# *WEEDING THE CROP*

## CULTIVATORS

As soon as a farmer had planted his seed, he began the age old battle against the weeds, which, if allowed to remain, would rob the soil and, in turn, the crop of needed moisture and nutrients. During the age of hand power, the most common way to kill weeds was to chop them out of the field with the hoe. These hoes were usually fashioned from iron by the village blacksmith. By 1823, however, at least two manufacturers in Philadelphia were making cast steel hoes, and in 1836, two Pittsburgh manufactures were making about 192,000 steel hoes annually, More than twenty years later hoe manufacturing was still a major aspect of the agricultural tool-making business, because in 1857, four Pittsburgh firms made 384,000 hoes. These hand tools were primarily used to cultivate corn in the North until about 1840, and for weeding cotton and corn in the South as late as the mid-1850s.

Corn farmers generally hoed their crop four times. At the rate of three-quarters to one acre per day, as much as six days labor per acre might be spent killing weeds by hand. Certainly, the amount of corn farmers could efficiently cultivate determined the acreage planted, and cultivation by hand severely limited the total corn acreage nationwide. About 1820, however, American farmers began using an implement called a "horse-hoe" to cultivate their crops. This horse-drawn equipment was based on a cultivator designed by Jethro Tull. Its purpose was to loosen the soil and kill the weeds. Although wheat farmers found it of limited use, corn farmers quickly began to see the cultivator's labor saving value, and a host of American innovations followed.

Those initial experiments with single-row, horse-drawn cultivators encouraged some farmers to try harrows and shallow turning plows for weeding between rows of corn. The shovel plow also became a favorite among corn farmers in both the North and South. The shovel plow was well suited for cultivation because it fit easily between corn rows, killed the weeds, stirred the soil, and tilled the plants by casting more earth around the newly sprouted corn. By 1850, a second shovel had been added and the double shovel plow became a com-

*The P.P. & Mast Company of Springfield, Ohio, manufactured the No. 4 Buckeye Sunbeam Walking Cultivator. This implement was designed with a high frame and an arched axle to pass over waist-high corn plants.*

mon cultivating tool among corn farmers.

In the mid-1820s, the expandable cultivator appeared. This implement consisted of a triangular shaped frame which could be expanded from twelve to twenty-eight inches in order to till between varying spaced rows. The cast-iron teeth only cultivated one row at a time, and they had a tendency to catch on roots, rocks, and stumps. Although this implement clogged easily, it cultivated more efficiently than the hoe, and it was lighter than a plow thereby improving maneuverability. Corn farmers believed they could do more work with this implement than with three plows. Consequently, this cultivator was an important advance in the evolution of cultivating implements.

During the 1830s, most farmers continued to cultivate with the hoe. Indeed, few farmers had ever seen a cultivator. Still, by the latter part of the decade more and more farmers were adopting the harrow or shovel plow for cultivation. Farmers commonly removed the front tooth on the triangular harrow in order to allow it to straddle the plant row. This method was usually used for the first cultivation. Subsequent tillage, where implements were used such as on the large corn farms in New York, Pennsylvania and Delaware, was done with the shovel plow. Cultivation techniques, however, differed from region to region, among localities, or even between farms depending on tradition and personal preferences. Small farms in New England, for example, which had only a few acres in corn, could be cultivated fairly quickly with the hoe. In contrast, plantations in the South with a large forced labor supply could send a number of workers into the cotton fields to cultivate by hand. However, farmers beyond the Appalachians, who had large acres of corn and wheat but a short supply of hired labor, turned more quickly to the newly developed horse-drawn cultivators. Even so, the shovel plow remained the common cultivating tool until the late 1840s and into the 1850s in Iowa and Illinois.

By the late 1840s, cultivators had been fitted with steel teeth which were less likely to break than cast-iron ones. Cultivators with iron teeth, however, were most common during that decade, particularly if the implements had five or six teeth. By the 1850s, iron-tooth cultivators were in common use in the East and Midwest, and steel-tooth cultivators were gaining in popularity.

During the 1840s and early 1850s, as well, wheat farmers in New York and Ohio began using the field cultivator. This large, horse-drawn implement was not used for cultivation at all, rather it was employed for cross plowing the seedbed prior to planting time. The field cultivator had a varying number of shovels, two

*Combination riding and walking cultivators were popular during the late nineteenth century. On this model, the levers regulated the depth of the shovels. When the farmer rode the cultivator, he used his feet to guide the shovels down the crop row.*

large wheels, a heavy frame and a seat. Somewhat smaller, seatless models, such as Ide's Wheel Cultivator, were also marketed for a short time. The cutting depth of these cultivators could be regulated by a lever or by screws on the axle. Few farmers in the Midwest cross plowed their lands, but these cultivators were used in this region in place of the harrow to cover wheat seed after using the broadcast seeder. By the late 1850s, the field cultivator was seldom used.

The next stage in cultivator development came in 1856, when George Esterly of Heart Prairie, Wisconsin, patented a new implement design. Although Esterly is incorrectly given the credit for patenting the first sulky cultivator at that time, he did patent a walking, straddle-row cultivator with increased flexibility and maneuverability. Esterly's cultivator consisted of two large wheels on an axle which was attached to a tongue. An eye bolt attached to each side of the axle from which a trailing beam joined a horizontal bar with movable shovels. These shovels were adjustable to expand or contract in order to cultivate on each side of the corn row. The farmer guided the cultivator with a set of handles attached to the rear bar.

During the 1850s, other innovations followed Esterly's. Some of those improvements involved arching the axle to straddle growing rows of corn and thereby enable the cultivation of more mature plants. Some changes involved the addition of shields to protect the plants, levers to raise and lower the shovels, or hinges to allow the shovels to spring or slip over obstructions. Two-row cultivators began to appear as well as implements without tongues. The tongueless cultivator became popular, because it had a light draft, it could be easily maneuvered close to the plants, and it could be turned easier at the end of each row. Farm implement manufacturers also added wrought-iron beams to cultivators during the 1850s. By the 1860s, the sulky cultivator was also in widespread use in the corn growing region of the Midwest. This two-horse implement cultivated on each side of the plants as one horse and one wheel traveled down a row while the other horse and wheel went down another. By cultivating both sides of the row at one time, corn farmers doubled the amount of land they could weed in a day, and they could ride at the same time. No cultivator, however, had greater than two-row capacity and most farmers preferred single-row implements.

In the *ante-bellum* South, farmers sometimes used the same types of cultivators that were being adopted among corn growers in the North. More frequently, however, if they did not use the hoe or the shovel plow, they cultivated between their tobacco and cotton plants

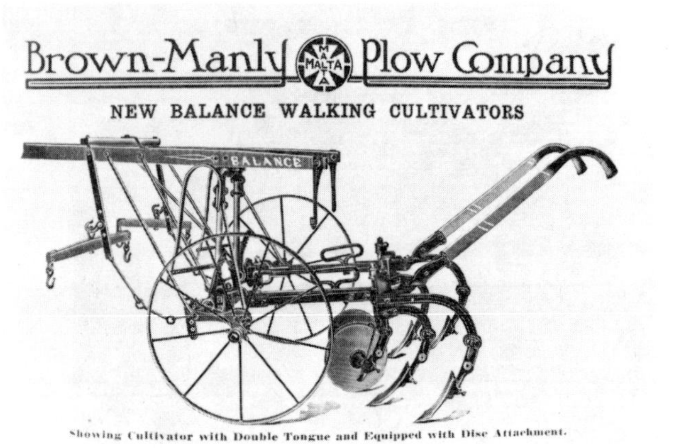


*Some cultivators were fitted with disks which tilled the soil close to the plants and helped the farmer guide the implement down the crop row.*

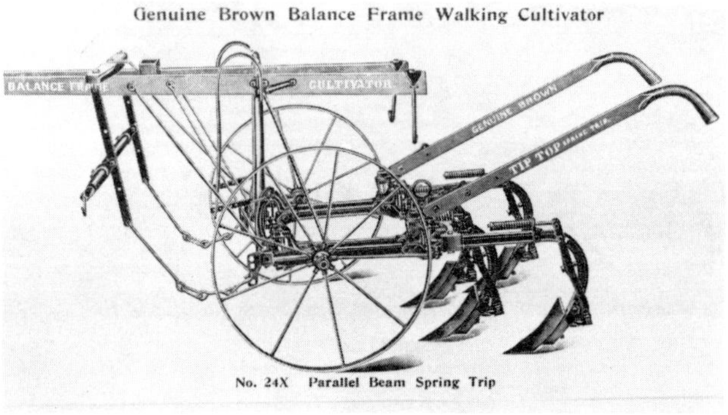


*The springs on this cultivator enabled each shovel to kick backwards without breaking, if it hit an obstruction.*

with implements called scrapers, skimmers, or sweeps. Generally, these were used only on the most progressive plantations. Scrapers were developed originally to cut weeds and loosen soil around tobacco plants without breaking the leaves. These cultivators had long, horizontal blades, made from thin pieces of wrought-iron, which shaved off the weeds a few inches below the surface. Scrapers came in various styles, but all of them cut a swath varying from approximately twelve to twenty-two inches wide. These cultivators were particularly well suited for level fields and loamy soil rather than for rough lands and heavy soils. Consequently, scrapers were more commonly found in Mississippi, Louisiana, Arkansas and in the black prairie lands of Alabama rather than on Georgia and South Carolina farms and plantations where hills and clay soils prevailed. With these implements, a farmer could cultivate about one and a quarter acres per day. In the North, a slightly different variation of this cultivator was fitted with steel plates called "duck feet." The duck feet attached to the beam and pared or cut the weeds like the scraper did in the South.

By 1870, the design of the cultivator had become nearly standardized with most manufacturers producing models with rectangular frames mounted on two wheels with a tongue and a driver's seat. For the remainder of the nineteenth century cultivator patents generally involved changes in detail. Those changes were known as "combination claims," that is, the inventors did not claim a new invention, but rather unique arrangements of the various cultivator parts. Indeed, 1,900 such patents were made in 1869 alone. Those claims often involved designing more efficient means for expanding or contracting the beam so as to adapt the implement to the width of the crop row. Other changes involved improving the shape of the cutting teeth or shovels in order to make them more adjustable or more efficient in throwing the soil in a certain manner or direction. Some cultivators had wheels, others did not. Some cultivators had springs attached to the shanks which allowed the shovel to trip backward, if it hit a solid object such as a rock. Once the obstruction had been cleared the shovel sprang back into place. Many cultivators simply had wooden pins which kept the shovels from bending backwards as they cut through the soil. If a shovel hit an obstruction, however, the force of the blow would break the wooden pin, thereby, enabling the shovel to fall backward and lift up and over the object. Before proceeding, though, the farmer had to reset the shovel at the proper angle and insert a new pin in place of the broken one.

In 1870, two-row cultivators, drawn by three or four horses, easily weeded fifteen acres per day, with (as one report indicated), "almost as much ease and comfort as a day's journey in a buggy." Sulky cultivators with four bottoms were also being used during that decade for cross plowing corn land in the spring. In this respect, they were used much like the old field cultivators had been some forty years before. For this purpose, though, a fifth shovel was often added at the front of the cultivator to give it a wedge shape which enabled it to cut through the soil more effectively. By the turn of the twentieth century, cultivator teeth and shovels were being made from soft-center steel for maximum durability. Disk cultivators were also being used where large amounts of soil had to be moved on the furrow, or when weeds were exceptionally high.

For the next fifty years after 1870, however, the rate of cultivation by horse-drawn implements remained about the same. Indeed, cultivation speed did not change remarkably until after 1924, when the International Harvester Company introduced the first affordable and efficient row-crop tractor. Tractors increased the draft power available which enabled farm implement manufacturers to add more rows of cultivator teeth or shovels to further speed the weeding process.

During the age of horse power, then, the cultivator

*By the turn of the twentieth century, two-row, sulky cultivators were popular among farmers with large cornfields. Note the wooden break pin behind the right-hand shovel. If the shovel hit a solid object, such as a rock, the pin broke and the shovel sprang back without breaking. Then, the farmer had to stop and reset a new pin. This procedure took time, but it was cheaper than replacing broken shovels. (Smithsonian Institution.)*

became, perhaps, second in importance only to the plow among the farmer's tools. Indeed, without the cultivator vast acreages of corn would not have been possible, because the weeds would have choked the crop before the farmer could hoe them out and bring in the harvest. When farmers adopted the cultivator in place of the hoe, they could plant larger crops because they now had the ability to weed more land more quickly and more efficiently than ever before. Still, change comes slowly and it depends on the willingness of individuals, such as farmers, to accept new tools and methods for the job at hand. Nevertheless, the cultivators which developed during the nineteenth century saved many man-hours in the field during weeding time, and when their worth was clearly seen, farmers readily adopted this most useful implement.

With the crops weeded, the American farmer prepared for harvest time. During this season too, new technological innovations made this task quicker and easier.

# CHAPTER V

# THE GRAIN HARVESTERS

Until the rapid adoption of horse-drawn machinery in the mid-nineteenth century, the American farmer harvested his wheat, oats, barley, and rye by hand. This was hard work which had to be completed swiftly, because once the grain was ripe, a race began with time and the weather. If the grain was not cut promptly, it might shrivel in the head; or, hail, wind, or heavy rain might destroy it. In the vast prairie land west of the Appalachians, wheat acreage was limited only by the amount a farmer could effectively harvest. This restriction contrasted with the corn crop which was limited only by the amount a farmer could cultivate, since the harvesting requirements of these two crops are entirely different.

For harvesting small grain crops, the sickle was the most common reaping tool until the late eighteenth century. The sickle is a ancient agricultural implement that has remained virtually unchanged since Mesopotamian farmers fashioned it from clay as early as 3,700 B.C. With its curved, metal blade and wooden handle, the sickle enabled early American farmers to reap the harvest by an age old method. To use this tool, the reaper stooped down, separated a handful of grain stalks with the point of the sickle and grasped them with his left hand. He then drew the sickle through the stalks, from the heel of the blade to the point, with his right hand. An axiom of that age was that no one learned how to reap properly until they had cut the little finger on the hand which held the bunched stalks. The stalks, once cut, would then be raked together, bound into bundles and laid together in shocks by helpers who followed the reapers through the field or by the reapers themselves. The sickle was well suited for rough, stumpy land or where the grain had fallen down and had tangled. Still, harvesting grain with a sickle was back-breaking drudgery. At best, a reaper could harvest only three-fourths to one acre per day with the sickle, depending on the thickness of the grain.

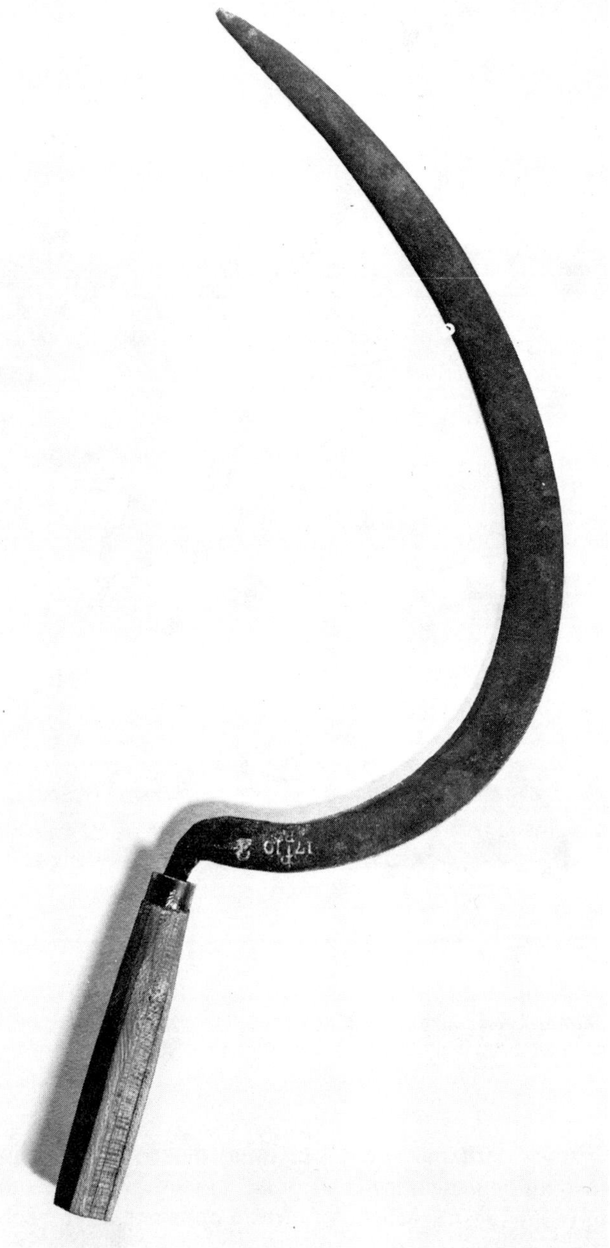

*Until the late eighteenth century, farmers used the sickle to harvest their grain crops. The sickle has remained virtually unchanged for nearly 6,000 years, except that it is now made from iron instead of clay. (Smithsonian Institution.)*

By the end of the War of Independence, however, farmers in the middle Atlantic states, particularly Virginia, Maryland, and Pennsylvania, were using a more efficient tool — the cradle scythe — for the grain harvest. Although the scythe was a European invention which had been introduced into the American colonies early in the seventeenth century for hay mowing, the American farmer changed its design in order to make it more suitable for cutting grain. The cradle scythe was unique. It consisted of a grass scythe secured to a frame with four or five long wooden fingers. This frame attached above the blade and the wooden fingers ran parallel to it. As the reaper made his cut, the grain fell onto the fingers. The reaper then tilted the cradle and allowed the grain to fall into a pile where it could be raked into a bundle and bound into sheaves. A skilled reaper could cut approximately two to three acres per day — approximately triple the amount he could cut with a sickle. The cradle scythe enabled the reaper to maintain an upright stance, but the ten- to twelve-pound tool required skill to manipulate, and cradlers commonly received more pay than did the other harvest hands. Still, once the grain had been cut, it had to be raked and bound for shocking. This involved stoop labor similar to that of cutting with a sickle. Generally, three binders for every two cradlers were expected to prepare about 1,000 sheaves per day for shocking over a ten-acre field.

Although, the cradle scythe achieved quick popularity among middle Atlantic farmers, it was seldom used in the South before the early nineteenth century. Southern farmers preferred the sickle, because it did not shatter the grain from the heads as much as the cradle scythe, and because it was easier to manipulate in heavy stands of grain. Many southern farmers also preferred the sickle, because it left more straw or stubble to be plowed under to help fertilize the field as well as to lessen threshing and stacking time, since there was less straw to work with. Many Pennsylvania and Ohio farmers also continued to use the sickle well into the 1820s, and New England farmers used this hand tool until the late 1830s. Still, the cradle scythe was adopted by more and more farmers, and it became a standard reaping tool until about 1860.

Whether grain farmers used the sickle or the grain cradle to bring in the harvest, these tools restricted the acreage they could expect to safely harvest in the course of the season, unless they were prepared to expend large sums of money to hire harvest hands. In the case of wheat, for example, a farmer might have a maximum of ten days to complete the harvest before it began to shatter out of the heads or even less time if a weather change threatened to slow or ruin the harvest. Until the farmer could speed the harvesting process mechanically, he had little hope of cheaply expanding his grain production. With land cheap and labor expensive, the cost of hiring a large number of harvest hands was often prohibitive, because many men and women preferred to own or rent their own farms rather than work for someone else. Only horse-drawn machinery would free the grain farmer from a dependence on hand tools and hired labor. But, until an efficient horse-drawn harvester of some sort was invented and perfected, the individual farmer's grain production was severely limited.

*In order to harvest grain with a sickle, the reaper had to bend down, grab a handful of stalks with one hand and draw the sickle through the grain with the other. Harvesting with a sickle was back-breaking work. (Smithsonian Institution.)*

*Mid-Atlantic states grain farmers were using the cradle scythe in place of the sickle by the end of the American Revolution. The cradle scythe enabled the reaper to harvest grain without bending over. (Smithsonian Institution.)*

## REAPERS

Gallic farmers had experimented with a form of reaper during the first century A.D. which, when pushed, stripped the grain heads from the stalks and caught them in a box. Many centuries passed before any other significant attempt was made to mechanize the grain harvest, and it was not until the eighteenth century that a mechanical reaper became practical. The first American patent for a mechanized reaper was issued to Richard French and T. J. Hawkins of New Jersey on 17 May 1803. The design of that machine is not entirely clear. It had three wheels, one of which extended into the grain at one side. The cutters consisted of a series of scythe-like knives which revolved on a vertical spindle. Long, wooden fingers extended into the grain below the cutter. A team of horses drew the machine from the side. Beyond this description, little is known, since the patent records have been destroyed. Nevertheless, American inventors had begun

*Above: The cradle scythe deposited the cut grain in a pile ready for binding. A skilled worker could cut three times as much grain with a cradle scythe as with a sickle. (International Harvester.)*

*Left: In 1834, Cyrus McCormick patented his first reaper after three years of experimentation. This reaper is a reproduction of McCormick's original invention. (Smithsonian Institution.)*

to direct their attention to the grain harvest. Other experiments followed which met with either failure or very limited success, and a new age in grain harvesting did not begin until 1831, when Cyrus Hall McCormick tested his first reaper in Rockbridge County, Virginia.

McCormick's reaper had a straight, knife-like blade, which was linked to the drive wheel by a pitman and gearing. The blade oscillated or reciprocated and sawed through the stalks as the machine moved forward. Projecting fingers or guards on the cutter bar caught and held the stalks while the blade cut through them. The grain fell onto a platform and was raked off by someone walking alongside. This method of clearing the platform kept the grain out of the way as the reaper made the next round, since the horses were harnessed to the side and followed in the previously cut path. A divider on the outer edge of the cutter bar separated the standing grain from the swath being cut.

McCormick was not entirely satisfied when he tested this machine in a rye field during the summer of 1831. After the initial test, he linked a reel to the main axle with a belt to help gather the grain in front of the blade. He also improved the divider and added saw-tooth serrations to the blade to improve its cutting ability. McCormick tested his reaper again in a field of oats where it successfully cut six or seven acres. McCormick spent the next three years trying to make further improvements before patenting the machine on 21 June 1834. This reaper was substantially the same as his 1831 test model. Even so, McCormick continued to make adjustments on his reaper and he did not place the machine on the market until 1840. In the meantime, on 2 July 1833, Obed Hussey, tested a reaper near Carthage, Ohio, before the Hamilton County Agricultural Society.

Hussey patented his reaper on 31 December 1833, and sold his first machines in New York and Illinois the following year. Hussey's reaper differed from McCormick's in several respects. First, it did not have a reel to help gather and hold the grain while the sickle bar cut it. Second, the five-foot sickle consisted of a series of triangular steel plates which were riveted to a flat iron bar. This cutter bar had a reciprocal motion between slotted, spike-like fingers. As the machine moved forward, the sickle clipped or chopped through the stalks. Hussey's reaper also consisted of a heavy frame which carried the gearing and the platform. The platform extended from behind the main wheels off to the right side

SENIOR HAND-RAKE REAPER SIDE DELIVERY.

*Prior to the development of a self-rake reaper, during the 1850s, someone had to stand or sit on the reaper or walk along beside it to rake the cut grain from the platform. The Russell & Company of Massillon, Ohio, offered the Senior Hand-Rake Reaper to the public for the 1869 harvest.*

HUSSEY'S REAPER, PATENTED IN 1833.

A. Platform which receives the grain. B. Tongue to which the horses are attached

*Above & Left: In 1833, Obed Hussey patented the first reaper which achieved modest commercial success. Hussey's reaper, shown here with a drawing and model, was pulled by two horses. Like McCormick's first reaper, the driver had to sit on one of the horses. (Smithsonian Institution.)*

*By 1841, Hussey had added a seat for the person who raked the cut grain from the platform. The single drive-wheel was added in 1837. (Smithsonian Institution.)*

of the machine, and it was supported by two small wheels. Two large drive-wheels carried the frame. The sickle was attached to a pitman which was given its motion by a crank geared to the axle. The grain fell onto the platform where it remained until enough accumulated to make a bundle. Then, it was raked onto the ground for binding.

By 1837, Hussey had changed the construction of the reaper by substituting one large ground- or drive-wheel for the original two. He still did not provide a seat on the reaper so the driver had to ride one of the two horses needed to pull the machine. By 1841, Hussey had added a seat and had moved the platform from behind the main wheel to the side. Hussey's machine, however, had a heavier draft than McCormick's and it had to be drawn much faster than the horses would normally walk — almost at a trot.

Each machine had its ardent supporters, but Hussey's reaper was best suited for mowing hay. This was not readily apparent, and both Hussey and McCormick entered into a long and bitter period of competition which did not end until Hussey sold his patents to rival manufacturers in 1858 and retired from the reaper business. Hussey's machine, however, was the only practical reaper sold during the 1830s. Although he built and sold about forty-five machines across the country by the time McCormick offered his machine for sale, it still required many improvements. Hussey's reaper, at first, for example, deposited the cut grain directly behind the machine so that it had to be raked, gathered, bound, and shocked before the machine made the next round, otherwise, the horses would trample it.

In 1843, a comparative test in Virginia between McCormick's and Hussey's reapers revealed positive and negative results for both machines. At that time, McCormick's reaper still had a lighter draft and required only two horses to pull it while Hussey's machine required four draft horses. McCormick's machine cut better in damp grain, but Hussey's reaper performed more satisfactorly in grain which had tangled or had fallen down. Both machines cut cleanly, but Hussey's cut approximately one-fourth to one-third more per day. Neither machine cut properly on hilly or uneven land. Still, if a field of wheat or oats was smooth, and clear of obstructions, most farmers who were familiar with both machines believed reapers would quickly pay for themselves with the cost of labor saved.

McCormick and Hussey continued to make improvements. In 1847, Hussey modified the cutter bar to eliminate the problem of clogging in wet grain. In that same year, McCormick changed his design to place the gearing in front of the drive-wheel to protect it from dirt and to give the reaper better balance. Two years later (1849), McCormick took the driver off the horse and gave him a seat on the reaper. And he further improved his machine about 1850, when he bought the rights to use Hussey's cutter bar on his machines.

In 1847 and 1848, Hussey's and McCormick's reaper patents expired respectively and they were not reissued. Other inventors now came forth with their own ideas for the improvement of the machine and made plans for its manufacture. By 1852, the reaper was no longer an experimental implement. Although additional improvements would be made over the next ten years, farmers could now confidently purchase the reaper and receive consistent, efficient service for about ten years before it wore out. Ohio, New York, and Illinois quickly became major reaper-producing states.

Sales now increased dramatically. Midwestern farmers had lands well suited for growing wheat and for mechanizing the harvest. With the reaper, grain farmers could harvest larger crops with less hired labor than ever before. Since the reaper did not shatter the grain out of the head as did the cradle scythe, less grain was wasted. The reaper also cleanly cut the stalks close to the ground and thereby increased the amount of straw saved over that produced by the cradle scythe. These features convinced many farmers that a reaper would pay for itself in one year. By 1851, McCormick was producing more than 1,000 reapers annually, and in 1852 an estimated 3,500 new reapers replaced 17,500 harvest hands in the Old Northwest.

During the Crimean War (1853-1856) European demands for wheat increased. At the same time the demand rose, prices climbed, the California gold rush of 1849 and the lure of the Far West continued to drain farm workers from the East and Midwest, so farmers began to meet increased grain demands and labor shortages by adopting the reaper. The McCormick, Manny, Ketchum, and Atkins reapers became popular implements during the 1850s. By 1855, the reaper was a common sight in the wheat fields at harvest time; and, by 1860 over 80,000 of these machines were operating on the grain farms west of the Appalachians. By the outbreak of the Civil War in 1861, an estimated seventy percent of the wheat in the West was harvested with the

# KIRBY'S AMERICAN HARVESTER.

For the general working of the Machine, we refer to the following gentlemen:

Henry Fitzhugh, Esq., Sherwood Forest, Stafford Co., Va.; L. C. Combs, Esq., St. Mary's River, St. Mary's Co., Md.; John F. Fenwick, Esq., Leonardstown, do. do.; Wm. H. Ford, Esq., St. Inigoes, do. do.; Chas. S. Somerville, Monticello, Calvert Co., Md.; W. W. W. Bowie, Esq., Prince George's Co., Md.; Hon. Richard B. Carmichael, Queen Ann's Co., Md.

PRICE OF COMBINED MACHINE IN BALTIMORE......................$140
Do. MOWER ONLY do. do. ...................... 120
Do. MOWER WITH REEL, do. do. ...................... 130

**Manufactured by D. M. OSBORNE & CO.,**

mar-3t 145 PRATT STREET, (OPPOSITE MALTBY HOUSE,) BALTIMORE, MD.

*Advertisement for Kirby's American Harvester. From **The American Farmer-Advertizer**, May, 1860. (Smithsonian Institution.)*

*From **The American Farmer-Advertizer**, June 1860. (Smithsonian Institution.)*

reaper. Wartime labor shortages further encouraged grain farmers to adopt this machine. In 1864, for example, an estimated 60,000 to 85,000 reapers were produced and sold — more than the total number manufactured between 1833 and 1861. In 1865, McCormick alone sold as many as 7,000 reapers. Four years later, 60,000 reapers were produced nationwide. At that time, reapers required a crew of 8 to 10 men, one to drive the horse, a second to rake the platform, and 6 to 8 men for binding and shocking. This crew could harvest from 10 to 12 acres per day.

Still, the reapers simply cut grain. In this respect, the reaper eliminated the need to hire cradlers and it speeded the harvest, but hired hands were needed to rake the gavels of cut grain together, bind the sheaves, and place them in shocks. Even though a man could ride on the reaper and rake the grain off, this too was tiring work. Consequently, by the mid-1840s, inventors were turning their attention to the development of an automatic raking mechanism which would remove the cut grain from the platform. Various techniques were tried to provide revolving raking arms. Many of those designs did not work very well and frequently scattered the grain off the platform, thereby making the raking and binding more difficult. Since these machines were more technically complicated than the hand-rake reapers, they were more difficult to maintain in a state of good repair. These early self-rake reapers were also more expensive than the hand-rake models, and they were not readily adaptable for mowing hay as were the hand-rake models. All of these factors, then, prevented quick adoption of these machines.

By 1854, however, the first commercially successful self-rake reaper, based on an 1851 patent, was marketed by the Brockport, New York, firm of Seymour and Morgan. The reaper called the "NewYork Self-Raking Reaper," or simply the "New Yorker," had a rake operated by a gearing from the ground wheel. As the

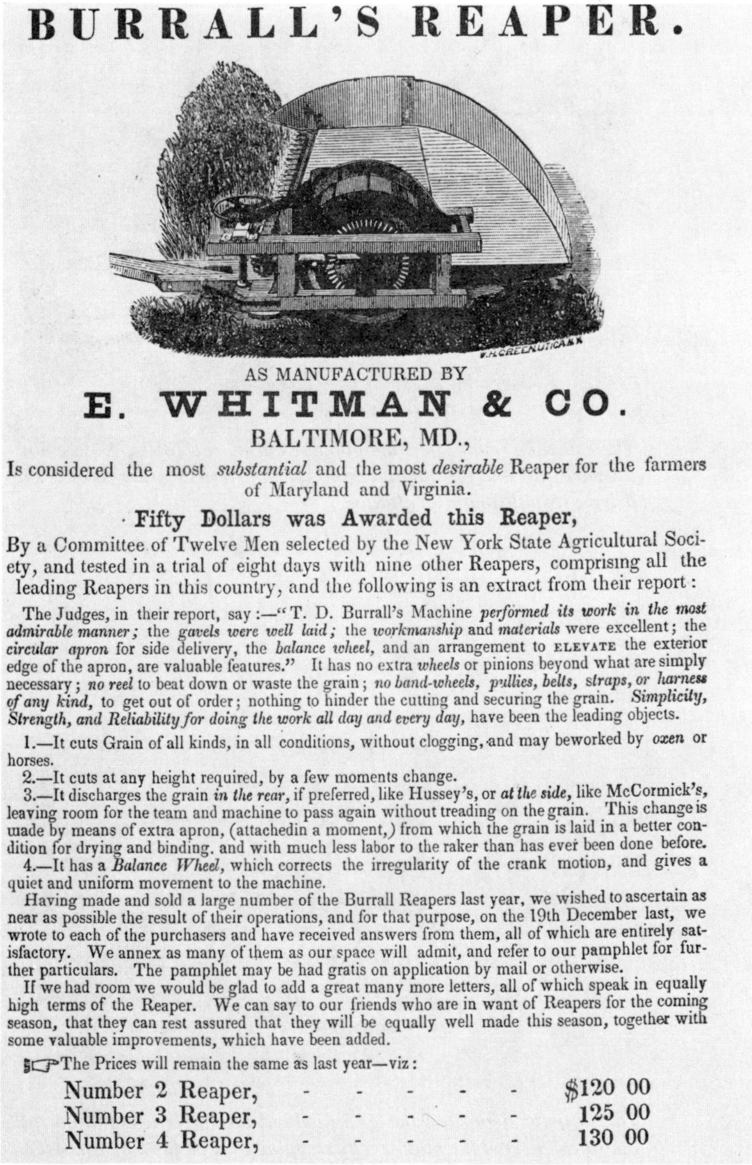


*From **The American Farmer-Advertizer**, May, 1855. (Smithsonian Institution.)*

machine moved forward, the rake swept across the platform at intervals and deposited the gavel on the ground where it awaited the binders. The wooden platform had a metal guard on the outer edge to prevent the grain from being brushed to the ground prematurely as the rake made its ninety degree arc. With this invention, one more worker was eliminated from the harvesting process.

Other manufacturing firms developed similar self-raking reapers. Owen Dorsey, a Maryland inventor, patented a popular design for a self-rake reaper in 1856. Dorsey's design allowed the rakes to revolve around a vertical axis by means of a cam. As the rakes rotated, they lowered at the front of the platform, swept across it, and deposited the gavel off to the side on the ground. After clearing the platform, the rakes rose and swung around to the front to begin a new sweep. At first, the rakes made too wide an arc to permit the driver to ride on the machine. This problem was eliminated in 1861, when changes were made in the gearing to lift and swing the rakes out of the driver's way. Machines with this raking mechanism were called "pigeon wing" reapers, and they became standard reaper features for the remainder of the nineteenth century.

A simplification of the self-rake reaper was known as the "dropper." The cutter bar, reel, and platform of this implement were identical to the reaper. The major difference, however, was that the hinged platform behind the cutter bar dropped at the rear and deposited the grain on the ground. In contrast to the reaper, the dropper did not deposit the gavel in a neat pile ready for binding. It simply dumped the grain from the platform onto the ground behind the machine. As a result, the binders had to work fast to clear the way before the next pass so the horses would not trample it. Although an 1849 patent granted to Oliver Barr of Illinois suggested the dropping technique for a reaper, the machine did not become important until after 1869, when Amos Rank of

*The Dorsey Reaping Machine became one of the early successful self-rake reapers. The rake arms rested on the curved metal hoop. The hoop provided support and served as a guide for the revolving arms. (Smithsonian Institution.)*

*The self-rake reaper became popular during the 1860s. Notice how the rakes lift and tilt after clearing the platform in order to miss the driver. (Smithsonian Institution.)*

WM. ANSON WOOD NUMBER ONE REAPER.

*In 1881, The William Anson Wood Mower and Reaper Company of Youngstown, Ohio, built this single-wheeled, self-rake reaper. This reaper cut a swath five feet wide, and it could be purchased with four or six rakes. A tool box was mounted on the beam.*

*The dropper deposited the grain behind the implement, when the operator lowered the platform by stepping on a pedal. The Russell & Company offered the Senior Dropper for sale in time for the 1869 harvest.*

*The Atkins Automaton or Self-Raking Reaper was built during the 1850s. The first Atkins reaper was produced in 1852, and, in 1856, 3,000 of these implements were manufactured for the harvest season.*

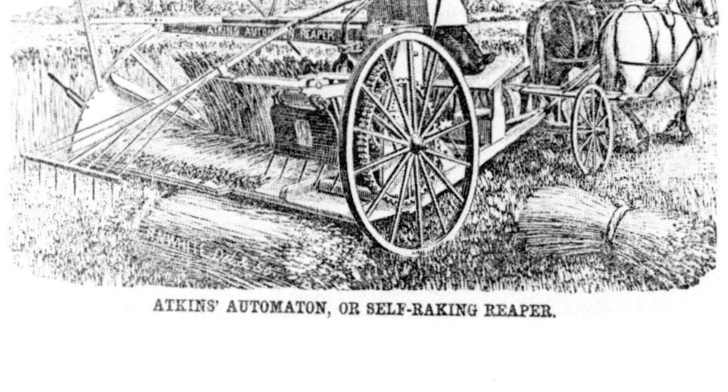

ATKINS' AUTOMATON, OR SELF-RAKING REAPER.

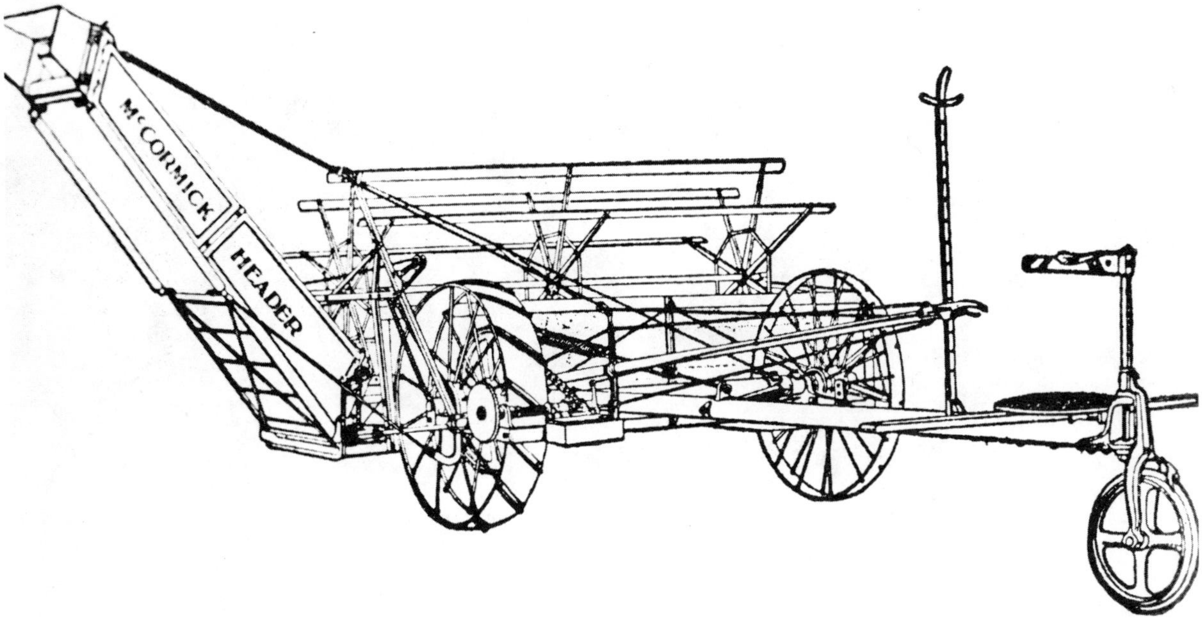


*During the late nineteenth and early twentieth centuries, many farmers in the Great Plains and Far West used the header to harvest the wheat crop. In contrast to the reaper, the horses were hitched behind the reel and cutter bar, and the operator steered the header with the tiller.*

*The Buckeye Dropper was manufactured by the C. Aultman & Company in Canton, Ohio. The dropper apparently did not deposit the gavel in as neat a pile as this illustration indicates, because farmers complained that it scattered the cut grain.*

Ohio obtained patents for the dropper and licensed several implement manufacturers to produce them. A number of companies manufactured the dropper during the later nineteenth century, but other grain harvesters worked better; and, on the large wheat farms of the Great Plains, they were seldom used.

In retrospect, by the outbreak of the Civil War, the major problems of the self-rake reapers had been overcome, and both production and sales increased. McCormick, for example, produced his first self-rake reaper in 1861; by 1864, two-thirds of his total reaper production consisted of self-rake models. Labor shortages during the war years made the self-rake reaper extremely popular, and the increased price of wheat made it more affordable than ever before. By the end of the war, where the reaper had reduced the number of men needed to harvest fifteen acres in a day from fourteen to nine hands, the self-raking reaper reduced that work force to eight men. Still, binders were needed no matter whether grain was raked off the platform by hand or by automation. Therefore, in order to eliminate the need for binders and to make the harvest faster and easier than ever before, inventors began to contemplate ways to eliminate hand binding from the harvesting process.

## HEADERS

During the late 1840s and early 1850s, another grain harvesting implement — the header — attracted attention in the midwestern wheat lands. On 2 October 1844, George Esterly of Heart Prairie, Wisconsin, patented the first practical header. At that time, Esterly believed Hussey's reaper required too much draft to become practical while McCormick's reaper was still of unproven merit. As a result, Esterly decided to develop a grain harvester which would strip the ears from the

*This McCormick header was used during the 1890s. To the far left, the driver stands on the platform behind the six-horse hitch. (National Archives.)*

*In order for these headers to cut clockwise without trampling the first swath of grain, the field was probably opened with a reaper or cradle scythe. Farmers with large wheat fields frequently cut from the center to the outside to avoid that initial opening around the edges of the field. (Smithsonian Institution.)*

stalks and collect them in a hopper. Esterly's header was not unlike the Gallic reaper used in the Roman Empire during the first century A.D. His machine had a straight, knife-like sickle which adjusted in height to slice the heads from the plants. A spiral wheel with paddles, powered by a belt attached to the ground wheel, revolved and forced the heads against the knife which sheared them off — much like a reel type lawn mower. The grain heads were then forced into a large bin. A four-horse team pushed this cumbersome, four-wheeled header through the grain field. The operator steered Esterly's header with a wheel, like that which a ship's captain used, while he stood on a platform in the rear.

Five years later (1849) Jonathan Haines of Illinois also patented a header. This was the only other header successfully produced before the Civil War. The "Haines Illinois Harvester" featured a reciprocating sickle (instead of a straight, fixed blade), and an endless conveyor to carry the grain heads to a wagon drawn alongside. The grain heads then only had to be hauled to the place where the threshing was to take place. Like Esterly's machine, the Haines header was pushed before a two-horse team. The operator steered from the rear with a tiller. The Haines Illinois Harvester cut a ten-foot swath and was advertised to cut from twenty-five to forty acres a day.

Although the header sold well in the prairie region

*The header wagons or barges hauled the grain to the stack for threshing. (Smithsonian Institution.)*

*This header was pushed by a hitch of three mules and two horses. (Kansas State Historical Society.)*

and in California from the late 1840s to the early 1860s, farmers soon found the grain had to be dry before cutting it with this machine; otherwise, it would spoil in the stack. Grain farmers in the humid Midwest seldom had the proper weather or climatic conditions to permit thorough ripening and drying before the grain was cut. Midwestern grain was generally cut while slightly green and it needed to cure in the shocks to permit adequate threshing. As a result, midwestern farmers rejected the header in favor of the reaper. During the 1860s, however, the header was the most important harvesting machine in California, and in the 1870s, it gained some popularity in Kansas and Nebraska.

By the last quarter of the nineteenth century, the header customarily had a 12-foot cutter bar and harvested from 15 to 25 acres per day. A four-horse team usually pushed the header and the operator carefully steered it close to the wagon which collected the heads from the machine's conveyor. In contrast to the reaper and binder, however, the header cut counter-clockwise around the grain field. Since the header was designed to lift the grain into the wagon from the left side of the machine, a counter-clockwise harvesting pattern was required to keep the horse-drawn wagon from knocking down the uncut grain as it followed alongside. Clockwise harvesting would have required opening the field with a binder or grain cradle to clear a path around the edges for the header and wagon. Instead, the header operator drove to the center of the field and began cutting in a circle from the inside toward the outer edges. The cut grain was then stacked in the center of the field where it awaited the threshing machine.

Overall, the header had several advantages over the reaper. The most important was that it eliminated binding, shocking and hauling the sheaves to the threshing site. Consequently, it reduced the number of workers required at harvest time. Since the header only cut the top eight to ten inches of the stalks, the threshing machine had greater capacity and speed than when grain cut by a reaper was fed into it. Nevertheless, the

MARSH HARVESTER—WARDER, MITCHELL & CO., SPRINGFIELD, OHIO.

*In 1864, the Marsh harvester reached the market. It delivered the cut grain to a platform where two men bound it into sheaves. Then, the bundles were tossed onto the ground ready for shocking.*

header was not well suited for small- or medium-sized fields, and it was somewhat cumbersome to steer. Consequently, it was soon replaced by the Marsh harvester and the automatic binder.

## HARVESTERS AND BINDERS

By the early 1850s, a number of inventors were trying to eliminate the backbreaking task of binding grain on the ground. All of those efforts failed to provide a solution to the problem until Charles W. and William Wallace Marsh of De Kalb County, Illinois, successfully tested and patented their harvester in 1858. The "Marsh harvester," though not offered for sale until 1864, consisted of a five-foot sickle and a reel similar to those used on reapers at that time. The machine differed from other reapers, however, because it had an endless canvas apron or belt which carried the cut grain away from the sickle and elevated it above the drive-wheel where it fell onto a receiving table. Next, two men riding on the attached platform gathered the grain and bound it into bundles which they dropped onto the ground. The Marsh harvester, then eliminated the stoop labor required for binding the sheaves. By 1870, the Marsh brothers had sold licenses for producing their machine to a number of manufacturers, and at least 1,000 were sold. A decade later, in 1879, an estimated 100,000 harvesters had been manufactured. Although, this machine cut 8 to 10 acres a day, mechanical breakdowns severely limited its popularity, many of the best self-rake reapers having a far superior performance to the Marsh harvester. Still, two men binding on the harvester could do as much work as four or five binders walking behind a reaper. Certainly, the Marsh harvester speeded the harvest, and while the machine did not become the complete solution to the farmer's harvest problems that the Marsh brothers had hoped, it was an important step toward the complete mechanization of the grain harvest.

Indeed, the first wire-binding mechanism, which eliminated the need to prepare the bundles by hand, was attached to the Marsh-type harvester in the early 1870s. Experiments had been conducted for the development of an automatic binder in 1850, when John E. Heath of Warren, Ohio, patented the first twine grain binder. Heath built several binders which worked fairly well, but he sold his patent rights in 1851 and made no further contributions to binder invention. Other innovations followed. At first, attempts were made to develop a reaper which would bind the sheaves with twine, but the lack of a suitable mechanism to tie a knot retarded the perfection of this binder. Instead, inventors found

*The binder automatically tied the cut grain into sheaves with wire or twine. This rear view of a binder shows the bundle carrier on the right. The bundle carrier caught the bound sheaves which fell from the tying mechanism above. When enough grain had collected to make a shock, the operator tripped a lever which emptied the platform. The canister held a spool of twine. (National Archives.)*

*Farmers with large wheat acreages usually hired several binder operators to harvest the crop. This is a harvest scene in Kiowa County, Kansas, during the 1890s. (National Archives.)*

that wire could be wrapped around a sheaf, cut, and twisted automatically to hold the bundle together.

The first machine to bind with wire, was patented on 18 November 1856 by C. A. McPhitridge of St. Louis, Missouri. The binding mechanism on this machine consisted of an arm which wrapped the wire around the gavel when an attendant turned a handle to trip the mechanism. The mechanism also cut the wire, twisted the ends together, and deposited the bundle on the ground ready for shocking. The operator had the task of determining when the proper amount of grain had been delivered to the platform where the bundle or sheaf would be bound.

James F. and John H. Gordon of Rochester, New York, eliminated the need for an attendant to gauge the size of the bundles, when they patented a device known as a "packer" which automatically measured the gavels and packed them into bundles for binding. James Gordon had been experimenting with this concept as early as 1862, but he did not achieve success until 1872 when he attached his binding mechanism to a Marsh harvester. This innovation worked well enough for Gordon to patent his device, but the first commercially successful wire binder, also made in 1872, was not his. At that time, Sylvanus D. Locke of Janesville, Wisconsin, produced a binder which had an improved mechanism which twisted the wire to tie the band. Although Locke sold only three binders in 1873, by 1875 he had joined the Walter A. Wood Company and about 300 Wood-Locke binders were sold that year. McCormick began manufacturing the wire binder in 1874, four years later he was producing 5,000 self-binders annually. In the same year, 1878, the Deering Company also began production by building several thousand wire binders. As a result, an estimated 20,000 wire binders were used in the 1878 harvest.

By 1875, the wire or self-binder achieved popularity, because it eliminated the need for rakers and binders; only one man, the driver, was required to operate it. The wire binder was particularly popular on the bonanza wheat farms in the Red River Valley. By the end of the decade, it was standard equipment on those vast wheat farms. Still, the wire binder was not without its problems. The wire often caught in the machine's moving parts, and, once the sheaves had been broken for threshing, disposal was a nuisance, since it would neither rot nor burn. Sometimes the wire damaged the threshing machine or mixed with the straw or grain. If the straw was fed to cattle, and the wire ingested, it could kill the livestock. Millers also complained that bits of wire damaged their grinders and sometimes passed into the flour. Some millers paid less for grain that had been bound by wire binders. Because of these problems, grain farmers were beginning to give up using the wire binder by the late 1880s. Fortunately for them, they did not have to return to binding by hand either on the Marsh type harvester or behind a reaper. Just at the time when the wire binder was coming under increased criticism, a mechanism for tying a knot in twine was perfected and adapted to the harvester.

In 1858, John Appleby of Whitewater, Wisconsin, began experimenting with a "bird bill" knotter and rotating arm which wrapped the twine around the bun-

*Once the grain had been bound and shocked, it had to be hauled to the threshing site. (Smithsonian Institution.)*

*The loose piles of sheaves were deposited by the bundle carriers on the binders. The worker at the left-center is setting the bundles into shocks. (Smithsonian Institution.)*

*Binding and shocking wheat on a bonanza farm in the Dakota Territory.*

dle and tied a knot in it. Appleby did not patent his device until 1875. In the mean time, Jacob Behel of Rockford, Illinois, also began experimenting with methods to solve the knotting problem. In 1864, Behel patented his tying mechanism called a "hawk bill" or "billhook." This knotter was similar to Appleby's, because Behel wrapped the twine around the bundle with a curved, revolving arm. Of these two pioneers in the twine-knotter field, Appleby was the most successful in gaining the support he needed to produce binders with his tying mechanism. The Deering, Esterly, Excelsior, McCormick, Buckeye, Champion, Minneapolis Harvester, Osborne, and Wood binders all used Appleby's tying mechanism.

In 1880, with the knotting problem solved, the Deering Company placed 3,000 twine binders on the market. McCormick entered the twine-binder market the following year and by late summer of 1882, he had produced some 15,000 machines. By that time, twine had replaced wire in the binder industry. In fact, McCormick stopped making wire binders in 1883. During the mid-1880s, competition among farm machinery companies caused the price of twine binders to decline, the price of hemp also fell. Farmers now could afford to purchase this machine with greater ease than in the past. By the end of the decade, almost the entire wheat harvest was brought in with the twine binder.

These early binders, though, were heavy machines. McCormick's five- and six-foot models weighed almost 1,600 pounds, and they placed a great strain on the draft horses. In 1885, however, the D. M. Osborne and William Deering Companies began substituting steel in place of wood and iron for the frame. This innovation made the binders lighter and more maneuverable. McCormick followed suit in 1887. By the late 1880s, the only non-steel parts were the binder's tongue, reel slats, and canvas elevator. During the 1890s, additional improvements were made to lower the center of gravity, to place small wheels on the tongue in order to take the machine's weight off the necks of the horses, and to add a bundle carrier. This last improvement was a hinged table which collected the bound sheaves until the operator tripped the release which deposited them in a pile ready for the shockers. The bundle carrier saved the shockers many weary steps, since they no longer had to walk back and forth to pick up the bundles and place them in shocks. Instead, they built the shock where the bundles fell. As a result, the harvest not only became quicker, but it also became cheaper since fewer shockers were needed to set up the bundles.

In retrospect, all of the mechanical harvesting implements — reapers, headers, harvesters, and binders — saved time and labor expenses. They did not, however, increase substantially the amount of grain a farmer could cut in one day. The hand-rake reaper and the twine binder averaged about ten to twelve acres per day, but occasionally as much as fifteen to twenty acres per day depending on field conditions and the thickness of the crop stand. Between the adoption of the self-rake reaper and the binder, horses and even steam engines could not provide the speed necessary to increase substantially the total number of acres that could be harvested in one day. An increase in daily harvested acreage would not come until the twentieth century when

THE BUCKEYE FOLDING BINDER, WITH BUNDLE CARRIER.

*Buckeye binder with bundle carrier.*

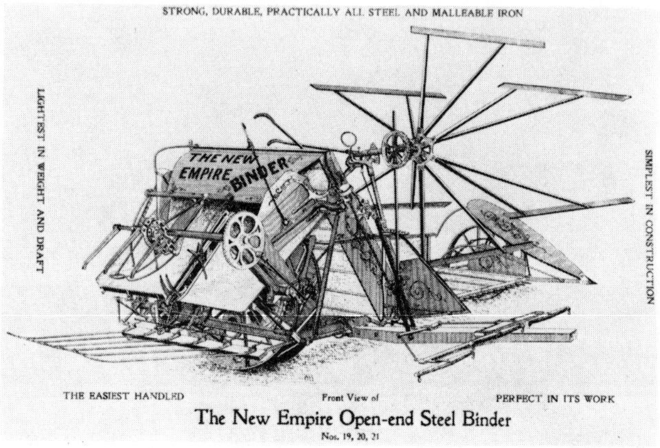


gasoline tractors provided the draft power necessary to expand the binder's daily cutting capacity. Still, these implements made the harvest substantially quicker, and easier for grain farmers. In the fifty year period between 1830 and 1880, the total time necessary to produce one bushel of wheat fell from three hours and forty minutes to ten minutes. In spite of improved tillage and seeding equipment, the tremendous savings in time was made possible primarily by the evolutionary change in grain harvesting technology.

Technological change in relation to the grain harvest, though, had another important effect besides saving the farmer time and money. As the development of the grain harvesters progressed, fewer and fewer hired hands were needed at harvest time. This change lightened the cooking burden on the farmer's wife who had the responsibility of preparing three hearty meals and frequently an afternoon lunch for the harvest hands from the time the cutting began until the last sheaf was in the shock.

During the last quarter of the nineteenth century, however, some farmers and inventors wanted to mechanize the grain harvest even more by combining the harvesting and threshing processes into one operation. By so doing, they hoped to eliminate entirely the need for shockers and threshing men. While the grain binders clattered across the wheat fields in ever increasing numbers, those changes were being made. Others wanted to apply the same kinds of technological change to the corn crop that had been so influential for harvesting small grains. During the late nineteenth century, several corn harvesting machines began to eliminate the task of picking corn by hand.

*Before the invention of the mechanical picker, the corn harvest was brought in by hand. Here, the ears are snapped from the stalks and tossed into the wagon. (Smithsonian Institution.)*

# CHAPTER VI

# THE CORN HARVESTERS

Of all the grains, corn not only produces the greatest yields, but before the age of the gasoline tractor, it also required more hand labor for harvesting than any other cereal. Corn stalks, laden with ears, are heavy and cumbersome to handle. Nevertheless, the application of technological innovation to corn harvesting lagged, because corn does not require immediate picking when it is ripe. If, for example, the farmer raised corn for the grain, the ears could be left on the stalk well into the winter. The dried stalks and cold weather did not damage the corn, and the harvest could be brought in after the other farm chores, such as haying and autumn plowing, had been completed.

When the corn was harvested, the work could be done in one of three ways. First, the corn stalks could be cut while still relatively green and placed in shocks. The ears would be removed at a later time and the stalks used for livestock feed. Second, the corn might be cut while it was still green and hauled to the barn for chopping into cattle feed. Or, third, the ears might be allowed to ripen fully in the field then picked from the stalks by hand and hauled to the barn or corn crib. The dry stalks were plowed under and allowed to decompose to help build soil humus.

No matter which harvesting method the corn farmer chose to employ, each method required a tremendous amount of hard work. Although the amount of time required to produce one bushel of corn fell from 4 hours and 34 minutes in 1855 to 41 minutes by 1894, this decrease was primarily attributable to the use of the gang plow and the check-row corn planter. The harvest was still brought in by hand.

Most corn farmers preferred to use both the stalks and the ears if at all possible. To do this, the corn was allowed to ripen until the ears were mature and fairly firm. At this stage of development, the stalks which were still somewhat green were cut with a sharp hoe or with long-bladed knives. The hoe was heavy to manipulate during the course of the day and most farmers preferred a corn knife. When using a corn knife, the harvester had to stoop at each plant and sever the stalk at ground level. This work tired the back, and most farmers either devised or purchased corn knives with long blades or handles which eliminated as much bending over as possible. Home-made corn knives were usually fashioned from old scythe blades. The scythe

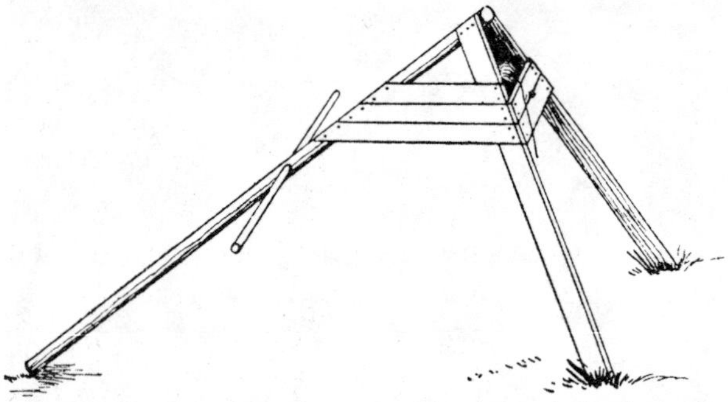

FIG. 2.—Wooden horse used to support shock.

*Some corn farmers used a wooden horse to build corn shocks. The bundles were set against the horizontal arm. When the shock had been completed, the arm was pulled out, the horse withdrawn, and the shock cinched tight.*

was cut into two parts and a corn knife hammered and filed from each piece. The knife made from the pointed end was usually the better of the two, because the knife forged from the scythe's shank was heavier to wield. A wooden handle was attached with screws or rivets. Factory-made corn knives came in all shapes and sizes. Still, corn harvesting with a knife was back-breaking work no matter which style of knife a farmer used. Some manufacturers attempted to eliminate entirely the need to stoop when cutting the stalks by producing a knife-like blade which attached to one's boot with a leather strap. Instead of bending over to cut the stalk, the farmer kicked the corn stalk with the blade to sever it. This idea resulted in just as much tiring work as when using a corn knife or hoe.

As the stalks were cut, they were gathered, bound into bundles and placed in shocks. Setting up the shocks required a degree of skill, otherwise they would collapse or fail to shed water properly and the corn might be damaged or destroyed by the weather. Shocks varied greatly in size and were usually made about as large as the individual farmer wished. Generally, a shock was made from an area twelve hills square or from 144 corn hills. The number of bundles tied also varied according to the height and thickness of the corn.

Corn shocks might be set up by several different methods. One common method was to use a long pole with crossing wooden arms inserted in it. The pole was shoved into the ground and the bundles leaned against it. When the shock approached the proper size, the rods were pulled out and the center pole withdrawn. The shock was then cinched tight with a rope and tied with twine or a cornstalk band. Some farmers used a three-legged brace, called a wooden horse, to help build corn shocks. One leg had a horizontal arm which the bundles or loose stalks were set against. When the shock reached the desired size, the rod was pulled out, the horse withdrawn, and the shock cinched tight. If a farmer preferred not to take a pole or wooden horse into the cornfield, he might build the shocks around a square of four bent-over corn stalks which had been twisted together. Once the corn was cut and shocked, the farmer still had to snap the ears from the stalk, husk them, and rebind and shock the stalks for later use, as well as haul the ears to the crib. This took time even though it could be done during the winter when other farm chores were less pressing. Many farmers recognized that if the corn harvest could be mechanized, the drudgery of handling the crop so many times could be reduced.

Actually, as early as 1820, an attempt had been made to mechanize the corn harvest. Other experiments followed, all of which met with either failure or very limited success largely because inventors were trying to apply the basic principles of the reaper to a corn harvester. The thickness of the stalks caused the early corn harvesters to clog or break down under the strain. While inventors tackled the problem of designing a mechanical harvester, some farmers began using corn sleds for cutting their crop.

The corn sled was simply a triangular, wooden platform with wooden runners or skids. A thin, sharp blade extended from both sides, and slanted forward from the rear. As the sled was drawn between the corn rows, the blades sliced the stalks which were then bound and set in shocks. Home-made corn sleds were probably used before the first one was patented in 1886 by J. C. Peterson of West Mansfield, Ohio. By the late 1880s, however, corn sleds were being manufactured and marketed by a number of implement companies.

The two-row sled reportedly cut enough corn in one day for 300 shocks. Two men rode on the platform and gathered the stalks in their arms to prevent the corn from falling in all directions after being cut. The sled stopped at each shock and loose stalks or bound bundles were added to it. This corn harvesting method was still hard work. The men riding the sled had to maintain their balance while sitting on a narrow seat, catch and hold the cut stalks, and walk back and forth between the sled and the shocks. Tangled corn stalks were hard to cut, and, even in the best corn, the horse had to work at a fast pace to permit proper cutting.

Still, the sled saved sufficient time and labor costs to warrant improvement. Stalk lifters or collectors were attached to the sled in order to guide the stalks onto the platform. When enough stalks had collected to make a shock, the workers riding on the sled pushed the stalks onto the ground for the binders who followed. Wheels were also added to reduce the draft and thereby lighten the work of the horse. And, knife guards were added to cover the cutting edges, when the sled was not in use in order to protect the legs of man and beast alike.

A more complicated version of the corn sled cut two

SCIENTIFIC HARVESTER NO. 1.
Showing wings set for In-cut. They can, however, be quickly changed to Out-cut if desired.

With Patent Automatic Knife Guards
Weight 285 Pounds. Code Word—Coharve.

*Some corn sleds or harvesters had reciprocating sickles to cut through the thick stalks and a horizontal arm to help build the shocks.*

*Top Left & Left: Corn sleds eliminated the task of cutting the crop by hand. One or two workers rode the sled and gathered the stalks in their arms until enough collected to tie a bundle. Or, they dropped the stalks in bunches to aid the binders who followed.*

rows at once with a reciprocating cutter bar. Although this design still attempted to imitate innovations in reaper technology, it worked efficiently enough to merit production by a number of implement companies. Two ground wheels powered the cutter bar. As the horse walked between the rows, the dividers, with endless guide chains on each side, channeled the stalks to the cutter. The stalks then fell onto a platform which tilted by tripping a lever. When a sufficient amount of corn had been collected on the platform, the horse was stopped, the platform lowered, and the shock set up around a pole which extended out from behind the machine.

Even though the corn sled was not the ultimate answer to speeding the corn harvest, it was an improvement over cutting the stalks by hand. Two men could cut and shock more than four and a half acres per day with the corn sled compared to less than one and a half acres per day for a man cutting with a corn knife. With the perfection of the twine binder for small grains, however, many farmers began anxiously awaiting the development of a similar machine that would ease the burden of the corn harvest.

## CORN BINDERS

During the 1880s, inventors began designing mechanical corn harvesters to help lessen the labor of cutting the stalks by hand. These one-row harvesters had endless chains on each row divider which fed the stalks into the cutter bar. The cut stalks fell upon an apron which elevated the stalks into a wagon drawn alongside the harvester. A binding attachment, however, could be substituted in place of the elevator. This machine did not work well enough to warrant widespread use.

In 1892, A. S. Peck of Geneva, Illinois, patented the first satisfactory corn binder. Peck's machine contained the essential principles used on all the corn binders produced thereafter. His machine consisted of two dividers which passed down each side of a corn row and fed the stalks into the cutter. The cutter bar was a serrated knife driven by gearing attached to the ground wheel. An endless chain caught the top and the bottom of the stalks and carried them back to the binding mechanism which packed and wrapped the bundle with twine before depositing it on the ground. The horses were hitched behind the binder in the same manner as they were on the header. Peck's corn binder worked better than any other implement, but two more years of experimentation were needed before the machine was suitable for manufacture and sale.

By the mid-1890s, the mechanical problems of the corn binder had been eliminated. Most farmers used the corn binder to harvest the crop while it was still green in order to run it through a shredder or fodder-chopper. The finely shredded stalks were then blown through an elevator and into the silo for ensilage. Most corn binders had row dividers attached to the frame. Two or

*Above: The Peck corn binder operated in front of a two-horse hitch. It was the first satisfactory corn binder.*

*Left: Some corn binders had special attachments which elevated the bundles into a wagon drawn alongside.*

*Below: By the mid-1880s, the mechanical problems of the corn binder had been eliminated. Generally, three horses were required for draft. (National Archives.)*

*This rear view of a corn binder shows the twine canister and a neatly tied bundle. (Smithsonian Institution.)*

three endless chains caught the stalks and carried them to a serrated cutting knife which was given a reciprocating motion by a pitman attached to a weighted flywheel. The flywheel gave the cutter enough force to slice through the toughest corn stalks. A conveyor carried the stalks back to the binding deck where the bundle was packed and tied with twine by a knotting device similar to that used on grain binders. Discharge arms then pushed the bundle onto the ground. These corn binders weighed an average of 1,500 pounds and worked best in fields where the corn had been planted in check-rows. The binder had a draft of 420 pounds and three horses were required to pull it. The corn binder harvested from seven to nine acres per day. Since the average field hand shocked about three and a third acres per day, two or three shockers were needed to follow behind the binder.

Although the cost of cutting an acre of corn with a binder was about the same as cutting it by hand and was slightly more expensive than by cutting it with a sled, (largely due to the cost of the binder and the expenditure for the twine), farmers still preferred the binder because it made the corn harvest easier. The binder, however, knocked some ears off the stalks which had to be picked up by hand after the corn was shocked. By the late 1890s, the implement companies were producing thousands of corn binders each year.

## SHOCKERS

Still, corn cut with a binder had to be shocked, and once the binder had been perfected, inventors turned their attention to devising a machine that would bind and shock the corn at the same time and thereby eliminate the need for additional field hands at harvest time.

Actually, the corn shocker was invented by A. N. Hadley in 1888. This machine consisted of a frame mounted on two wheels. Dividers with endless chains delivered the stalks to the reciprocating cutter. A five-foot wide circular table rotated behind the cutter. A shaft with radiating arms gathered the stalks in a vertical position and formed a shock on the table. When the table was full, after about 100 stalks or hills had been cut, the shock was hand tied and lifted onto the gound by the revolving crane mounted on the frame. The corn shocker harvested the corn crop cheaper than the sled or binder, and it built and set-up each shock in about five minutes time. The shocker only harvested about three or four acres per day, but it eliminated the need for two or three binders and therefore provided a savings in

The corn harvester and shocker eliminated the 2 or 3 men needed to follow the binder. The shocker platform held the stalks from about 100 hills.

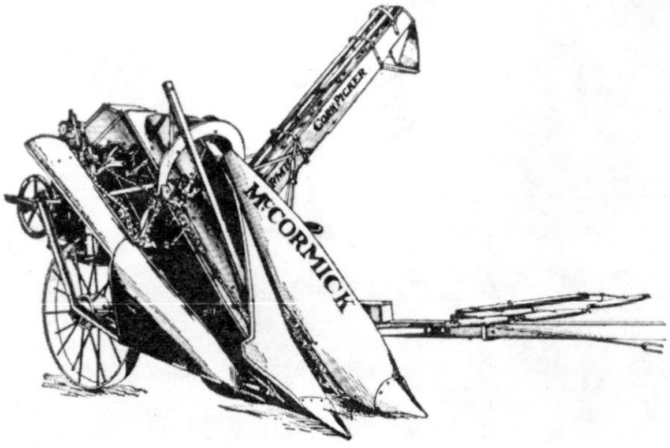


*The mechanical corn picker snapped the ears, stripped the husks, and elevated the cobs into a wagon drawn alongside.*

labor costs. Corn harvested with a shocker was more difficult to load into wagons than that cut with the binder. The corn bundles from the binder were light enough to load with a pitch fork, but the mechanically-made shock was so heavy a hoisting device had to be attached to the wagon. Even so, many farmers preferred the shocker over the corn binder because of the amount of labor saved.

## PICKERS

In the Midwest, farmers often raised more corn than they could use for fodder. Consequently, they did not need to shock the entire crop for livestock feed. Instead, midwestern farmers primarily raised corn for the ears which they picked by hand. This hard, monotonous task was accomplished in one of two ways. First, the corn farmer might snap the ears from the stalks while walking down the rows. The ears were tossed into piles and hauled to the crib later. Second, a high "bangboard" could be attached to a wagon on the side opposite the pickers. This board prevented the pickers from missing the wagon as they tossed the ears into it. A skilled picker could snap 100 bushels of ears per day. The wagon did not need a driver since the horses knew their business and walked ahead on their own.

In 1850, Edmund W. Quincy, of Peoria, Illinois, patented a corn picker which promised to bring an end to the task of snapping the ears from the stalks by hand. Quincy's machine was designed to pick the ears with two spike cylinders which tore off the ears as the corn stalks passed between them. The ears fell onto a conveyor which delivered them into a wagon driven alongside the machine. Although Quincy's implement never worked efficiently enough to merit manufacture and sale, he stimulated other inventors to try their efforts at solving the picking problem. A host of picking devices with rollers, cylinders, cutters, gathering prongs and parallel vibrating bars followed--all designed to snap or tear the ear from the stalk. These machines usually had row dividers to feed the stalks into the cutter. Generally, these early corn pickers were pushed by the horses hitched to the rear.

Manufacturers did not begin producing roller type corn pickers until 1874, even then, the machine was far from a state of perfection. Frequently, the rollers shelled the corn from the cob instead of snapping the ear from the stalk; or, some rollers failed to hold up under the strain of the thick, tough, corn stalks which passed between them. By the mid-1880s, roller-type corn pickers had been sufficiently strengthened consistently to snap the ears, strip the husks, and lift the ears into a wagon drawn alongside. But, the picker still had the disadvantage of shelling too much corn from the cob, and it failed to pick in tangled stalks. If the corn picker had a husking device attached, the draft increased substantially and four horses, instead of three, were required to pull it. If a husking attachment was not used, the rollers removed from 25 to 75 percent of the husks. The remainder had to be removed by hand or by a husker set up at the crib. Faced with these disadvan-

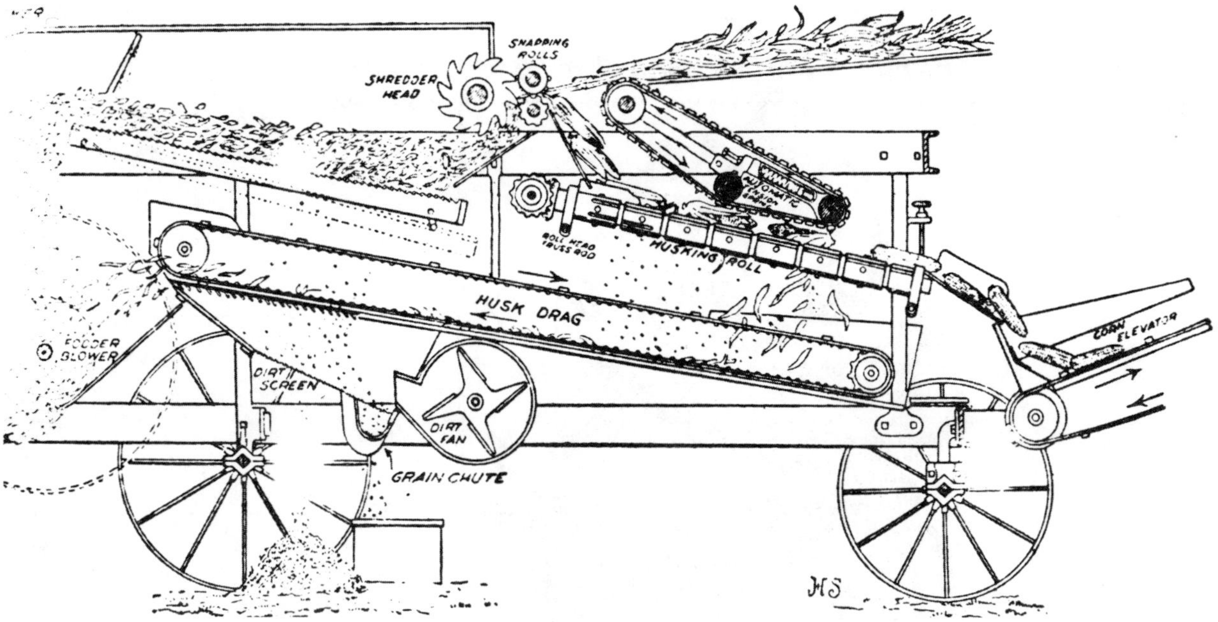


*The husker-shredder snapped the ears from the stalks, stripped the husks, and deposited the cob out one end while the fodder was blown out the rear. Any corn shelled in the process fell to the bottom of the machine and dropped out into the box on the ground.*

tages many farmers preferred to pick their corn by hand or use the binder or shocker. Farmers who used the picker, however, harvested seven to nine acres per day. During World War I, labor shortages and high harvesting costs forced many farmers to begin using the horse-drawn picker. Even by that time, though, the machine did not work very well, because the groundwheel still failed to provide sufficient power or consistent speed to operate the picking mechanism. The gasoline tractor, with its power take-off, would solve these problems, but it was not used for corn harvesting on a general basis until after the mid-1920s.

## HUSKERS

Whether farmers harvested their corn by cutting the stalks and placing them in shocks or by picking the ears, the husks still had to be removed. A variety of husking pegs and gloves were used for this task. The husking peg or glove had projecting iron spikes, which stripped the husk from the ear with greater ease than if the job was done with the hands alone. By using the husking peg, a farmer could husk about one acre of shocked corn per day or one and a half acres from uncut stalks in the field.

In 1837, Jonathan Cutler of Putney, Vermont made an attempt to mechanize the husking process, when he patented the first corn husker. That implement had a pair of rough rollers which revolved inward and stripped the husks off as the ears passed between them. Hand-powered huskers of this type were produced for sale by the mid-1860s. Some of those huskers were attached to a wagon so the ears could be husked while the picking was done. Prior to 1880, however, no machine had been invented that would satisfactorily husk the ears.

During the last decade of the nineteenth century, inventors gave increased attention to mechanical corn threshing or shelling. The implements tested at that time, however, so pulverized the stalks that inventors soon realized a mechanical shredder would have great appeal to farmers who used the stalks for forage. The finely-ground stalks were more palatable to cattle and livestock wasted less fodder if it were cut into fine pieces. This discovery led to the invention of the combined husker-shredder which removed the ears from the stalks, stripped the husks and ground the stalks into fodder.

During the 1890s, the husker-shredder gained popularity. Although many different models appeared on the market, the husker-shredders all worked about the same way. The stalks were fed into a roller which snapped off the ears. Once the ears were removed, the stalks were pushed through a series of knives attached to a spindle and pulverized. The shredded fodder then fell into a blower which passed it through a funnel and out of the machine onto the stack. The ears dropped from the snapping rollers to the husking rollers. When the husks were stripped away, the ears passed to an elevator which carried them out of the machine and into a bin or wagon at the opposite end from the fodder blower. The

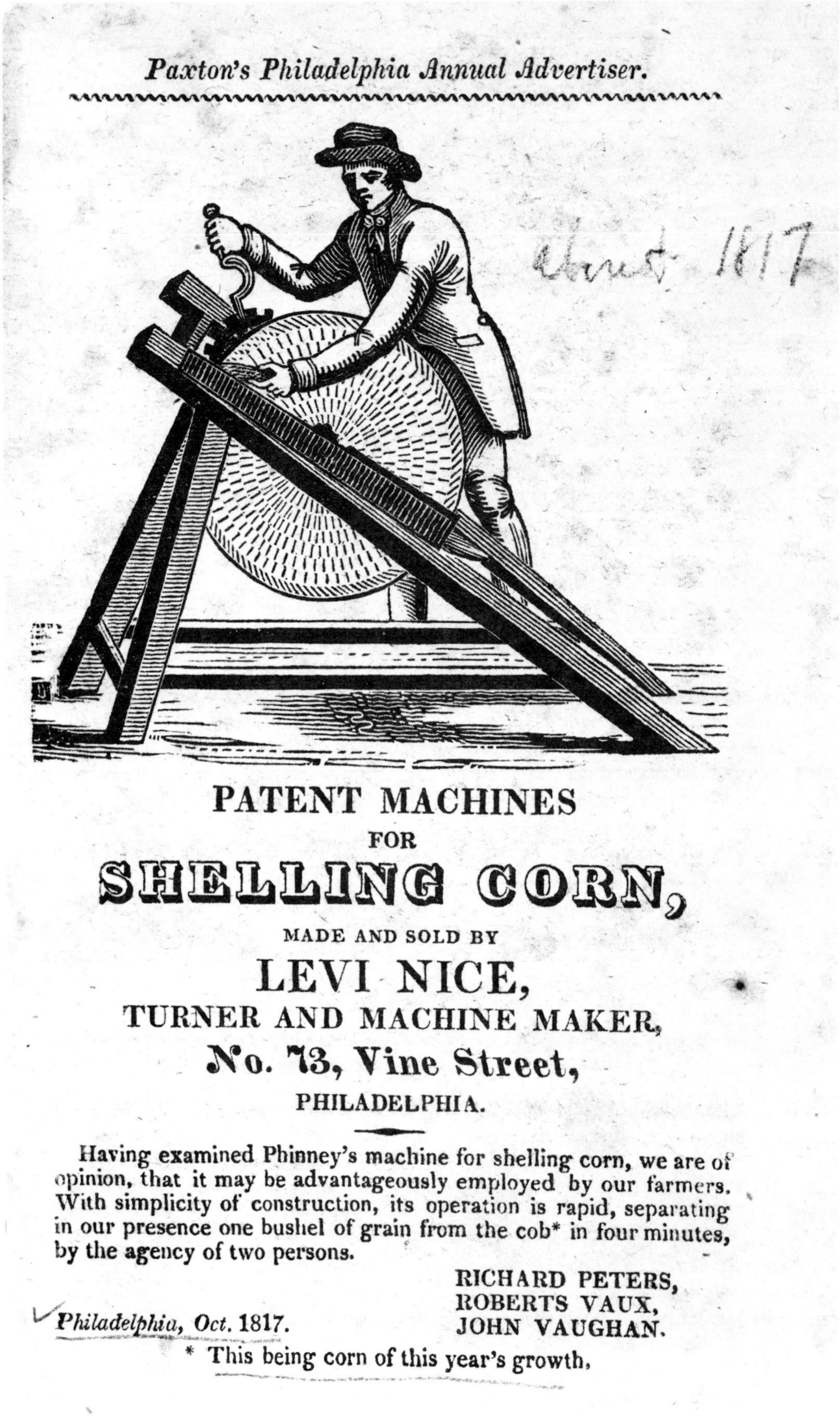


*Corn sheller advertisement. (Smithsonian Institution.)*

husks fell upon a conveyor which carried them back to the blower. Corn kernels which were removed by the husking rollers fell through a screen where a fan blew away dirt particles before the grain dropped into a box underneath the machine.

By the turn of the twentieth century implement manufacturers were producing husker-shredders which husked from 100 to 1,000 bushels per day. The larger machines were usually owned by custom operators. The smaller machines were more suitable for the farmer who preferred to do his own work since they were cheaper. At first, horses and steam engines provided the power to operate the husker-shredders. By the early twentieth century, however, many corn farmers were using small gasoline engines to drive the machine.

Southern farmers, however, did not readily adopt the husker-shredder during the nineteenth century. In fact, they seldom used the husking peg when picking corn in the field. Southern farmers preferred leaving the husks on the ears arguing the husks prevented insect damage while the corn was in the crib. Northern farmers, however, preferred removing the husks because they believed the husks attracted insects and rodents, took up too much crib space, and prevented proper drying of the ears. If the corn was not husked when it was picked, that job had to be completed prior to shelling.

## SHELLERS

Shelling is the process of removing the kernels from the cob. It was one of the most time-consuming aspects of the corn harvest. Early colonial farmers scraped sea shells along the ear to remove the corn and it is perhaps from this work that the term "shelling" is derived. Until the development of mechanical shellers a variety of methods were used to remove the corn from the cob such as flailing, treading with oxen or horses, scraping over an iron blade, and driving the ear through a metal ring with a mallet.

Prior to the 1820s, American farmers began using a corn sheller which consisted of a large, solid wheel with iron spikes on the surface. The wheel was mounted on a frame, and, as the operator turned the handle, the wheel revolved and the ear of corn was held against it. The

*Corn shellers stripped the kernels from the cob. The corn fell into a basket and the cobs dropped away through another opening. During the 1840s, the Burrall corn sheller was popular.*

kernels dropped onto a tray and the cob was either thrown away, burned or ground into fodder. Many variations of this sheller were placed on the market prior to the 1840s. Most of those shellers took the ear down a spout or iron throat which fed into a box where the spiked wheel removed the kernels. The kernels and the cob were generally deposited out the bottom together. By the early 1840s, shellers had been designed to drop the kernels out one spout while the cobs exited from another. The Burrall sheller was a popular model of that type. It was made from cast-iron and it had a fly wheel to equalize the velocity of the handcrank. One person turned the crank while another fed the cobs into the spout. The shelled corn fell out the bottom into a tub, and the cobs dropped out on top. With this implement about 40 bushels of corn could be shelled per day.

In 1843, F. N. Smith of Kinderhook, New York, patented the first horse-powered sheller. Smith's sheller, known as the "Cannon" was one of the best implements for shelling corn on a large scale. The "Cannon" sheller had an iron-tooth cylinder which was six feet long and fourteen inches wide. The ears were shoveled into one end of the cylinder, and, as the teeth raked the kernels from the cob, the corn dropped into a bin below the machine. The cobs exited from the end of the sheller. Two horses provided the power, and the machine shelled 100 bushels per day.

By the last quarter of the nineteenth century, two-horse shellers had been so improved that 2,500 bushels

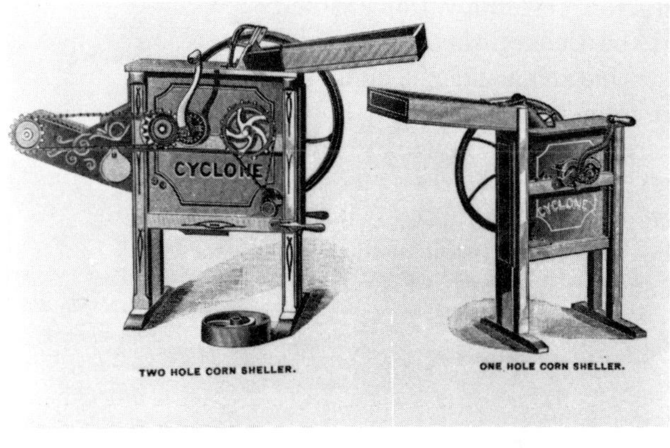


*Hand operated corn shellers.*

This corn sheller was powered by horses or by a steam engine. It was used to shell corn from large cribs. The cobs were fed into the machine at the right. The shelled corn was lifted out through the spout behind, and the cobs were taken away from the machine by the conveyor at the left. This implement could shell from 200 to 300 bushels per hour.

of corn could be shelled per day. By that time, most shellers had cast-iron or wood pulleys for attachment to a steam engine's belt. A self-feeding conveyor with iron teeth carried the ears into the shelling mechanism where beaters or corrugated wheels removed the kernels. A fan blew away the chaff and a cupped conveyor lifted the shelled corn into a wagon or bin. The cobs were carried out of the machine on a chain-lugged conveyor and dropped into a pile.

From the turn of the twentieth century until the 1920s, corn harvesting technology remained static. In the 1920s, however, the sheller and picker were combined and placed on a tractor. This innovation enabled the picking and shelling to be accomplished in one operation, a feat which horses and steam engines could not match. By the time corn farmers were adopting horse-drawn and steam-powered shellers, however, small grain farmers had already adopted threshing machines. These machines, clearly eliminated the arduous task of threshing grain with the flail, or the slow process of treading it from the heads with horses' hooves.

IRON SHELLER WITH TABLE, FAN, COB RAKE AND SACKER.

*In 1890, the Foos Manufacturing Company of Springfield, Ohio, manufactured this corn sheller. It could be powered by hand or by horses. This implement could shell from 250 to 300 bushels of corn per day.*

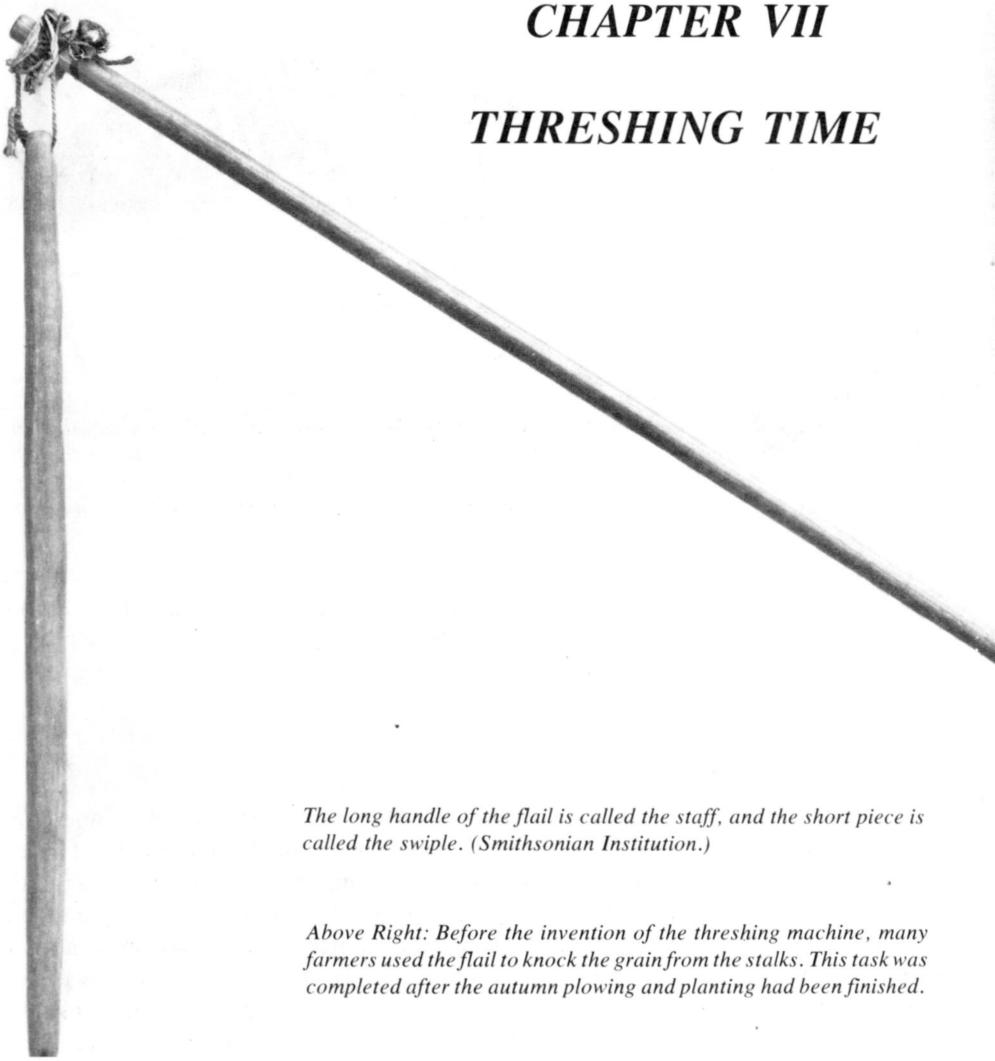

# CHAPTER VII

# THRESHING TIME

*The long handle of the flail is called the staff, and the short piece is called the swiple. (Smithsonian Institution.)*

*Above Right: Before the invention of the threshing machine, many farmers used the flail to knock the grain from the stalks. This task was completed after the autumn plowing and planting had been finished.*

After the grain had been harvested, the farmer had to thresh it. Threshing is the process of removing the grain from the heads. Once the grain was threshed, the next task was to winnow it — to seperate the grain from the straw and chaff. The jobs of threshing and winnowing were left for the winter months, and threshing could be accomplished in the barn and out of the weather, after the fields had become too wet or frozen for further plowing. Cold, dry weather was the best time for threshing, because if the grain was damp it did not beat out of the heads properly. Nevertheless, threshing and winnowing required long hours of hard work. As planting and harvesting machines increased the amount of grain a farmer could produce, technical innovations were necessary not only to speed the threshing time, but also to help the farmer cope with the increased harvest.

## FLAILS AND FANNING MILLS

Colonial farmers, particularly those in New England, used the flail to thresh the grain from the heads. The flail consisted of a short wooden club attached to a long handle with a piece of leather. The long handle, called the "staff," enabled the thresher to maintain an upright position as he lashed at the grain spread across the barn's threshing floor. The short, club-like piece, called a "swiple," struck the grain, and, in time, knocked it from the heads. Wielding the flail required considerable skill and beginners usually knocked themselves about the head and shoulders before they learned how to use the tool properly. After the grain was threshed, the straw was raked away and the grain and chaff swept into a pile or collected in a bin where it was stored for winnowing. On the average, approximately 7 bushels of wheat, 18 bushels of oats, 15 bushels of barley, 8 bushels of rye, and 20 bushels of buckwheat could be threshed and cleaned during the course of a ten-hour day. At this rate most of the winter would be needed to thresh the harvested crop.

In the middle or bread colonies, where grain production was greater than in New England, the flail was too slow to enable completion of threshing in a reasonable amount of time. There, in order to speed the threshing

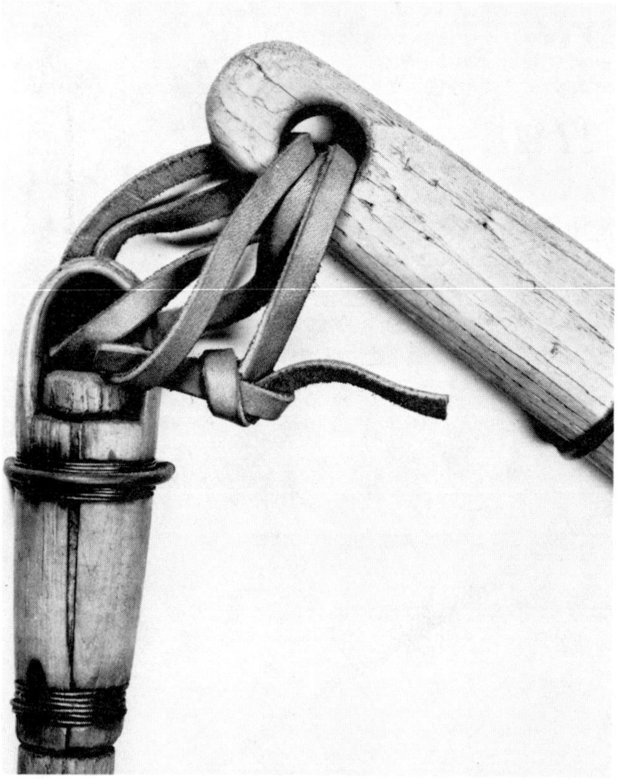

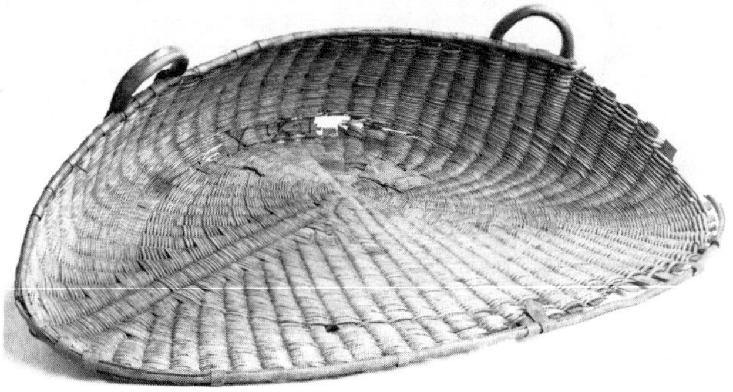

*Above: Once the grain had been threshed, it had to be cleaned. Many early American farmers used the winnowing basket to separate the grain from the chaff. In order to do so, they shook the basket's edge into the wind so the chaff would blow away. (Smithsonian Institution.)*

*Left: The staff and the swiple of the flail were usually joined by a leather strap. (Smithsonian Institution.)*

process and to keep the number of hired laborers to a minimum, farmers used oxen or horses to tread the grain from the heads. In order to do so, they spread several grain bundles in a circle on the wooden or stone threshing floor, or perhaps even on frozen or hard-packed ground, and walked the animals over it. A man stood in the center of the floor and guided the animals with the reins over the grain sheaves. The straw was frequently turned with a pitchfork so that as much of it as possible came into contact with the oxen's or horses' hooves. When the grain had been trampled from the ears, the straw was raked aside and the grain and chaff swept into a pile ready for cleaning.

Treading was substantially faster than threshing by flail. A man and a boy using three horses could tread about 30 bushels of grain per day, while two men and six horses could thresh approximately 100 bushels during that same period. Overall, however, threshing the grain from the heads with oxen or horses was not as efficient as flailing because the animals did not tread evenly and much of the grain was wasted. Consequently, the flail remained a common tool for threshing until about 1850. Nevertheless, if a farmer had a large acreage of small grain, such as wheat, to harvest, treading was more economical than flailing. The larger volume of grain produced and the speed of treading lowered labor costs because fewer workers were needed.

Once the grain was threshed, the farmer turned to the next task — winnowing the chaff from the grain. Several methods were used to separate the grain from the chaff. First, the straw was raked aside and the grain collected and placed in a wide, shallow basket and tossed into the air.The breeze blew the chaff and dust away, and the grain fell back into the basket. The farmer continued to toss the grain until as much chaff as possible had been removed. A second winnowing method involved pouring the grain from one basket into another until the chaff was blown away. If a breeze was not blowing, someone had to use a cloth sheet or blanket to fan the chaff from the grain. Some farmers preferred using a riddle to remove the chaff from the grain. This was simply a sieve which held the larger pieces of chaff as the grain fell through the screen onto the ground. The fine pieces of chaff which passed through the screen blew away to the side. Once the grain was winnowed, it was stored in bins or put into bags to facilitate handling and transport. Winnowing by hand remained common on farms which produced small grains well into the nineteenth century. Indeed, around 1820, one of Ohio's best wheat farmers winnowed his crop with a riddle while two persons fanned the chaff away with a sheet of cloth.

By the American Revolution the fanning mill had been introduced from Great Britain to speed the job of winnowing. The fanning mill consisted of a series of wooden paddles approximately 18 to 24 inches long which were attached to a rod geared to a crank. The paddles or fans were enclosed in a box-like frame which also housed several screens. As the grain and chaff was poured into the container at the top, the farmer turned the crank which caused the fans to take in air through apertures in the sides and blow it across the screens. The grain fell onto the screens and sifted to the bottom as the forced air blew the chaff away. The cleaned grain

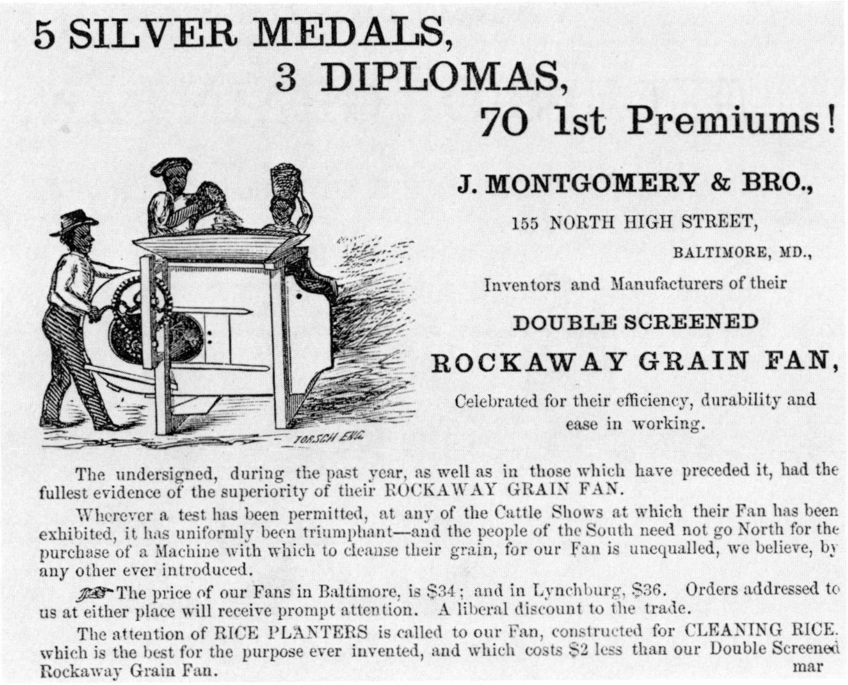


*Fanning mill advertisement from* ***The American Farmer-Advertizer****, July, 1854. (Smithsonian Institution.)*

poured out into a basket below. American-made fanning mills were available to the farmer soon after the turn of the nineteenth century. By the late 1830's, it had become a standard farm implement in the mid-Atlantic states and in the wheat producing region of New York.

By 1840, the winnowing basket, riddle, and breeze had been abandoned in favor of the fanning mill. The fanning mill remained popular among grain farmers through the nineteenth century even though the threshing machine had made it obsolete for large jobs by the 1840s. Many farmers used it to clean seed before placing it in the hopper of their grain drills. Seed cleaned in this manner did not clog the drill tubes, enabled more uniform planting, and permitted a better crop stand. The fanning mill required three men to operate it — one to turn the crank, one to pour the grain into the mill, and one to sack it and stack the grain bags. By using a fanning mill with a capacity of one bushel per minute, an average of from 400 to 600 bushels of grain could be cleaned and sacked in a ten-hour day. This average includes time for running the grain through the mill twice — the common practice to ensure proper cleaning.

## THRESHING MACHINES AND HORSE-POWERS

No matter whether a farmer threshed his grain with a flail and winnowed it with a basket or riddle, or whether he trod it from the ears under horses' hooves and blew the chaff away with a fanning mill, these tasks involved long hours of hard work. As early as 1788, however, Andrew Meikle, a Scottish inventor, patented a water-powered threshing machine which heralded the beginning of innovation to mechanize the threshing season. Meikle's machine did not winnow all of the straw and chaff from the grain, but other innovations quickly followed which improved the threshing and cleaning processes.

Satisfactory results from those experiments came slowly. In 1791, Samuel Mulliken, a Philadelphia inventor, patented the first threshing machine in the United States. This machine was too complicated to work efficiently and little advance was made among American inventors until the 1820s. Some American grain farmers imported British threshing machines after the War of 1812, but these machines were expensive, mechanically unreliable and in need of more horse-power than the average farmer could afford.

Between 1820 and 1830, however, a number of small, simple, inexpensive and locally-made hand- and horse-powered threshing machines appeared on the market. Jacob Pope, a Boston inventor, built the most popular thresher at that time. This machine had an endless belt which fed into a spiked cylinder and concave. As the grain was fed into the thresher, the cylinder with iron teeth rotated, and, together with the teeth fastened to the inside of the chamber or concave, beat the grain from the heads. These machines did not separate the straw or winnow the chaff from the grain, they simply threshed. Consequently, the grain and straw were deposited on the ground for separating and winnowing. Pope's machine threshed more efficiently than most other implements at that time, but farmers complained that it was harder to turn the crank than to wield a flail. To eliminate that problem, horse-powered gearings were soon added to Pope's and other manufactured threshing machines.

VAN WICKLE'S

CELEBRATED

**EXCELSIOR GRAIN FAN,**

MANUFACTURED BY

**J. A. BLACKBURN, 258 S. Caroline-st., Balto., Md.**

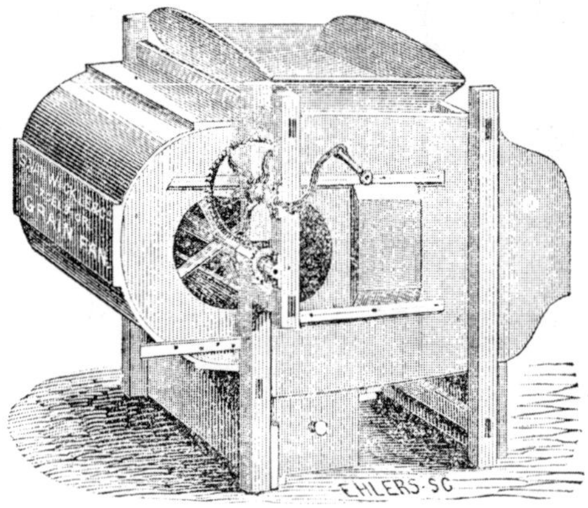


This superior GRAIN FAN has been sold for the last five or six years, in Maryland, Delaware and Virginia, and has made many warm friends, the ease with which it works, the perfect satisfaction it gives in chaffing and second cleaning, its simplicity of operation, its durability, &c., render it the most desirable Fan to Farmers generally, in the market.

Reference could be given to hundreds, but a few will suffice, to wit: Dr. Geo. R. Dennis, of Somerset Co., Md.; R. S. Todd, Esq., Snow Hill, Md.; Gen'l Walter Mitchell, Charles Co., Md.; Col. N. H. Shipley, Prince George's Co., Md.; Dr. R. S. Stewart, Anne Arundel Co., Md.

PRICE, $30—four months time.

28—Cash. SOLD BY

**THOMAS NORRIS,**

*Sole agent for Baltimore City.*

may-tf JOHN A. BLACKBURN, MANUFACTURER.

*By the American Revolution, farmers were using the fanning mill in place of the winnowing basket to clean grain. The fanning mill had become a standard farm implement by the late 1830's. (Smithsonian Institution.)*

By the early 1830s, approximately 700 threshing machines were on the market. These early implements were called "groundhog" threshers, because they were staked to the ground and had the appearance of digging into the earth while they were in operation. The groundhog threshers were driven by treadmills, sometimes referred to as railway horse-powers, and by horse-powered sweeps which were often called cider-mill horse-powers. The treadmill was the most common form of power on small farms. This device, consisting of an endless belt made from wooden slats, was mounted on rollers in an inclined position. The wooden belt and rollers were supported by a heavy frame. A fence-like pen was attached above the slats and rollers. One or two horses were placed in the pen, and, when the brake on the belt was released, the weight of the horses moved the slats backward and caused them to walk forward which, in turn, caused the belt to move continually. Since the treadmill depended on the weight of the horses for power, it generated only a relatively small amount of power.

Farmers with larger amounts of grain used bigger machines than could be adequately powered by a treadmill. In order to solve their power problem they adopted the horse-powered sweep. The sweep, in contrast to the treadmill, depended on the strength of the horses for its power. Consequently, the more horses a farmer could harness onto the sweep, the more threshing power he could obtain. The stationary sweep consisted of a center-post or spindle with radiating beams to which from two to eight horses were harnessed. As the horses walked in a circle, the spindle pivoted and turned a large master-wheel which was usually located above the spindle. As the master-wheel turned, it drove a pinion and shaft which transmitted power to the thresher by means of a belt or tumble-rod. During the 1840s, sweeps were mounted on wheels and fitted with folding booms in order to make them portable. Portable horse-powered sweeps eliminated the overhead master-wheel and placed the gearing at the bottom thereby forcing the horses to step over the tumble rod at each round.

In 1837, Hiram A. and John A. Pitts of Winthrop, Maine, improved the threshing machine when they patented one designed to thresh, separate the straw, and winnow the chaff from the grain in a single operation. A two-horse treadmill powered this small, portable thresher. As grain bundles were fed into the threshing cylinder, the grain was beaten from the heads, and an endless, vibrating, riddle-like conveyor belt, made from wooden slats, carried the grain and straw away from the cylinder. Most of the grain fell between the slats, and was winnowed by the fan which blew the chaff from the machine. The cleaned grain emptied into bags through a spout and the straw and chaff piled up at the rear of the machine and had to be carried away.

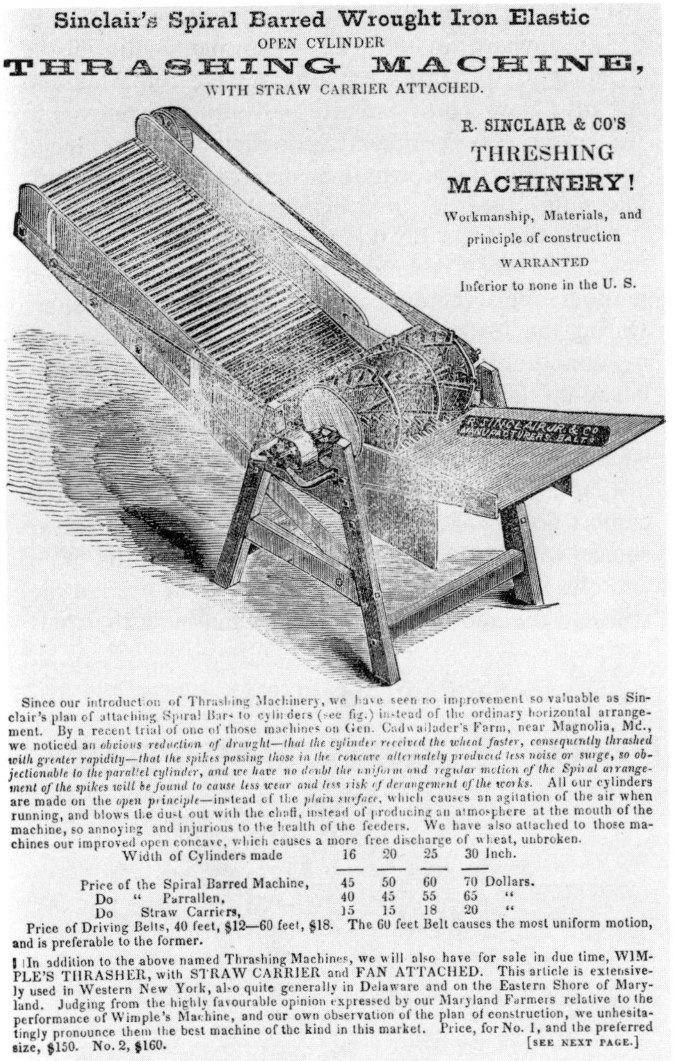


*Left: Between 1820 and 1830, the small, horse-powered threshing machine appeared on the market. On this machine, the stalks were fed into the spiked concave from the right, the threshed grain fell out below, and the straw was carried away on the conveyor at the left. From* ***The American Farmer-Advertizer****, May, 1856. (Smithsonian Institution.)*

*Below: During the mid-1820s, Pope's threshing machine was popular among New England farmers. This hand-operated machine did not separate the grain from the chaff or straw. (National Archives.)*

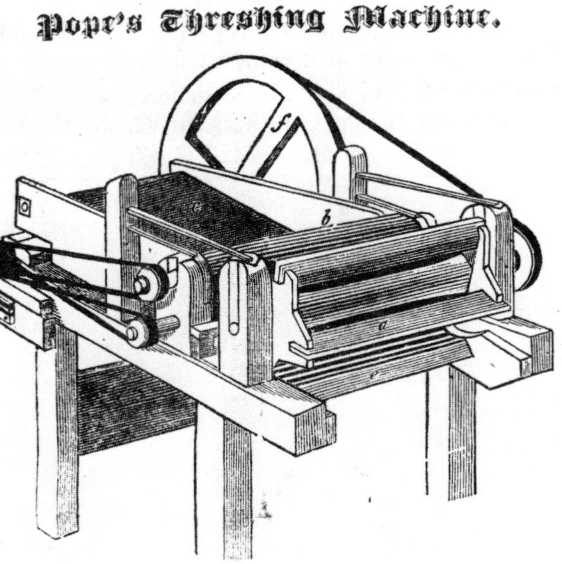

*This threshing machine was powered by horses. Notice the pulley on the left which a belt linked to a horse-powered treadmill. From* ***The American Farmer-Advertizer****, July, 1859. (Smithsonian Institution.)*

Although perfection of the threshing machine still had not been achieved, the Pitts thresher was a major advance in the threshing process. With the Pitts thresher, four men — one to deliver the bundles from the stack, one to feed the sheaves into the machine, one to bag the grain, and one to pitch the straw away from the thresher — could thresh about 100 bushels per day using a two-horse team. This was twice as much as they could thresh with the early groundhog machines. The Pitts thresher had the added advantage of light weight and small size. It could be loaded onto a two-horse wagon and transported to the place where it was needed. The thresher could be set up in about thirty minutes, and it could fit into a barn space as small as twelve-feet square.

During the 1840's the Pitts thresher became quite popular among wheat farmers. By the 1850s, mechanical threshers were in common use. Some machines could thresh from 300 to 500 bushels per day. By that time, however, the Pitts brothers had separated. About 1847, Hiram moved to Alton, Illinois, and then to Chicago where, in 1851, he began manufacturing the "Chicago Pitts" thresher. John moved from Maine to Albany, New York, and after several other moves to Rochester, New York, and Springfield, Ohio, settled in Buffalo, New York, where he manufactured the "Buffalo Pitts" thresher until his death in 1859.

After the Pitts brothers patented their threshing machine, other inventors and manufacturers began making improvements on the groundhog thresher. During the 1840s, for example, George Westinghouse manufactured a thresher at Schenectady, New York, based on the patent design of Jacob V. A. Wemple, a blacksmith and wagon-maker in Mineyville. This widely used implement had a threshing cylinder similar to that used on the Pitts machine, but it had a slatted canvas conveyor which vibrated as it revolved over square rollers. The vibrating motion shook the grain and chaff out of the straw. The conveyor carried the straw to the end of the machine, a fan blew the chaff

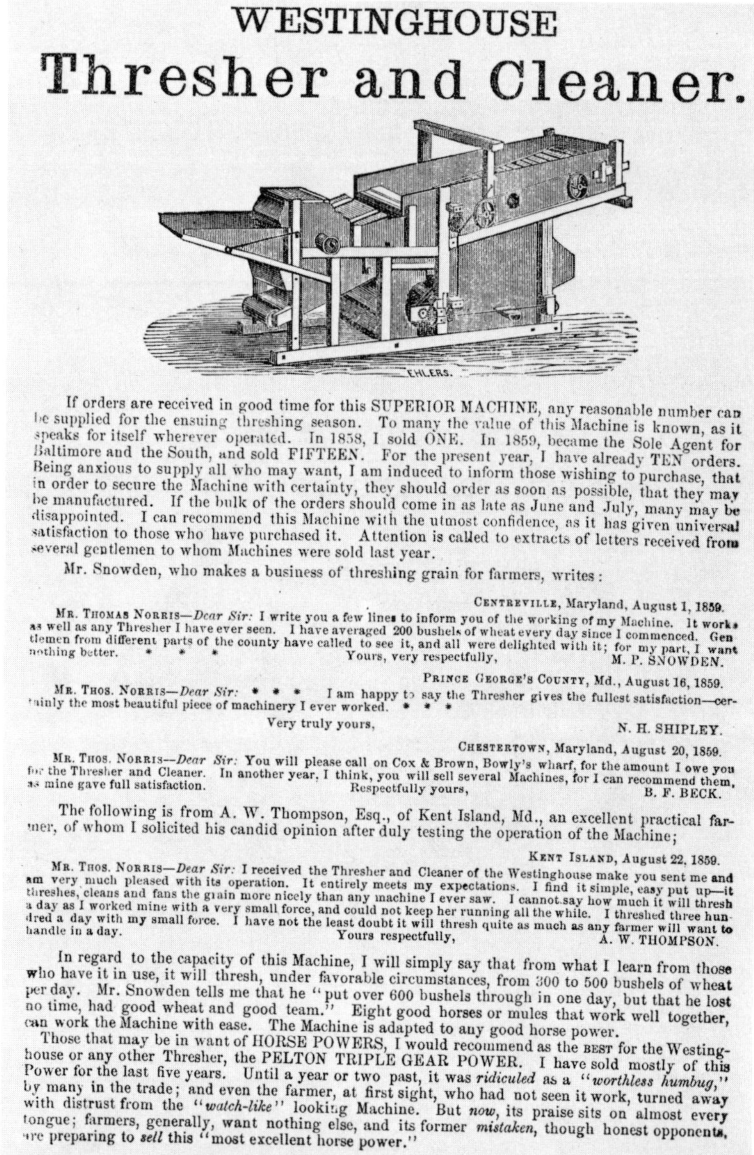


*The early threshing machines, such as this Westinghouse Thresher and Cleaner, were called "groundhog" threshers, because these implements tended to dig into the ground. (Smithsonian Institution.)*

*Above: On small farms, the treadmill most commonly powered the threshing machine. Power transferred from the treadmill to the threshing machine by the belt in the center of the illustration. (Smithsonian Institution.)*

E. WHITMAN & CO'S HORSE POWER.

The above cut represents a Horse Power that is well known in this country. It has been a long time in use, and is still highly approved, being simple, strong and durable. We are manufacturing four sizes, as follows:

No. 1.—Light 4 Horse Power.................................$80 00 | No. 3.—6 Horse Power.........................................$110 00
No. 2.— 4 do. ................................ 90 00 | No. 4.—8 or 10 Horse Power................................. 130 00

☞SEE THRESHING MACHINE ON ANOTHER PAGE.

*Left: Before the age of portable steam engines, horse-powered sweeps were needed to drive large threshing machines. The power transferred from the sweep to the machine by a tumble rod. That rod is beneath the cover at the left. (Smithsonian Institution.)*

away and clean grain emptied out ready for bagging.

During the late 1840s, John Cox and Cyrus Roberts of Belleville, Illinois, made additional changes or improvements in the groundhog thresher. Cox and Roberts replaced the endless wooden belt behind the cylinder with an agitating pan which had holes bored in the bottom. The grain and chaff fell through the holes as the pan vibrated and moved the straw away from the threshing cylinder. The grain and chaff dropped onto a ridddle where a fan blew the chaff away. The cleaned grain then passed through a spout into bags. This agitating principle was patented in 1852 with additional improvements patented in 1856. The St. Louis manufacturing firm of Kingsland and Ferguson first produced this machine in 1856, but the Battle Creek, Michigan, company of Nichols and Shepard began manufacturing the Cox and Roberts thresher in 1858. This company was primarily responsible for popularizing the "vibrator" thresher among grain farmers. The next year (1859) the vibrator thresher was redesigned to incorporate a double shaker or two vibrating troughs. The two-shaker design provided a counter balance and eliminated the tendency of threshing machines with only one trough to move along

*During the 1840s, implement manufacturers began building portable horse-powered sweeps. Portable sweeps gave the farmer greater flexibility at threshing time. These sweeps were characteristic of those built during the late nineteenth century.*

the ground while in operation.

During the late nineteenth century, farmers who did their own threshing usually owned small portable machines. The larger more efficient machines were generally owned by an entrepreneur who sent a thresher with an itinerent crew from farm to farm. These threshing-time entrepreneurs contracted their work for each season. Although contract threshing imposed an immediate labor expense on the farmer, it did free him from the capital investment necessary to purchase a large machine. By the Civil War, grain farmers commonly hired itinerent crews with threshing machines to do the work quickly. Large threshing machines saved time, and enabled many farmers to get their crop to market before prices fell.

## STACKERS AND FEEDERS

By the end of the Civil War only two major problems remained before the threshing machine could be truly claimed to have reached perfection. One involved the straw removal from the machine. Since the threshing machine simply dumped the straw on the ground, it had to be removed with pitchforks and piled in a stack. Pitching hay onto the stack involved an added labor expense for the farmer, and the men who worked behind the thresher had the dirtiest job of all. By 1870, however, most threshing machines had elevator attachments which carried the straw away from the machine and lifted it onto the stack. The stackers were generally endless conveyor belts made from chains or wooden slats. The conveyors were driven by the same power which operated the machine — either horse or steam. These stackers saved the expense of hiring at least two men.

During the 1880s, moveable or swinging straw stackers began replacing stationary elevators on the threshers. The swinging stacker moved from side to side and deposited the straw evenly on the stack. This stacker was not permanently attached to the thresher. Instead, it was mounted on wheels and moved independently of the thresher. The swinging stacker was linked to the thresher with a belt which was driven by the machine's power source. Although the swinging stacker worked better than the fixed stacker, it was clumsy and difficult to transport. In 1884, James Buchanan, an Indianapolis, Indiana, inventor, solved this problem with a blower. Straw blowers ultimately replaced the stationary and swinging stackers.

Buchanan's ''wind stacker'' consisted of a fan and a steel tube. As the fan turned, it created a vacuum which drew the chaff and straw out of the threshing chamber and into the tube which conveyed it to the stack. The stacker oscillated and adjusted in height and thereby enabled the threshermen to build high, wide stacks, easier and more quickly than ever before. In contrast, to the swinging stacker, the wind stacker was permanently attached to the thresher. The major disadvantage of the wind stacker was that it added from $250 to $300 to the cost of the threshing machine. Because of that extra expense, many farmers continued using threshing machines with stationary or swinging straw stackers until well after the turn of the twentieth century. Nevertheless, the wind stacker was far more efficient than any other stacker. It blew the chaff nearly fifty feet into the air and replaced as many as six men from this aspect of the threshing process. With the wind stacker, only one or two men were needed to build the straw stacks.

At the same time inventors were working out the problems of the hay stacker, they also turned their attention to developing automatic feeding and band cutting devices. The early threshing machines required two men to stand in front of the machine to cut the twine or wire from around the bundles before the stalks could be fed into the machine. During the 1880s another endless belt was added to the thresher to help feed the grain into the threshing cylinder. And, by the turn of the twentieth century, reciprocating knives had been placed at the front of the threshing cylinder to chop the twine bands into pieces so they would not clog the threshing machine.

During the 1830s and 1840s, many farmers complained that threshing machines were not only too expensive for the average farmer, but that they also ruined the grain for seed and the straw for feed. Some farmers also argued that threshing with a flail was cheaper than investing in a threshing machine, because the work

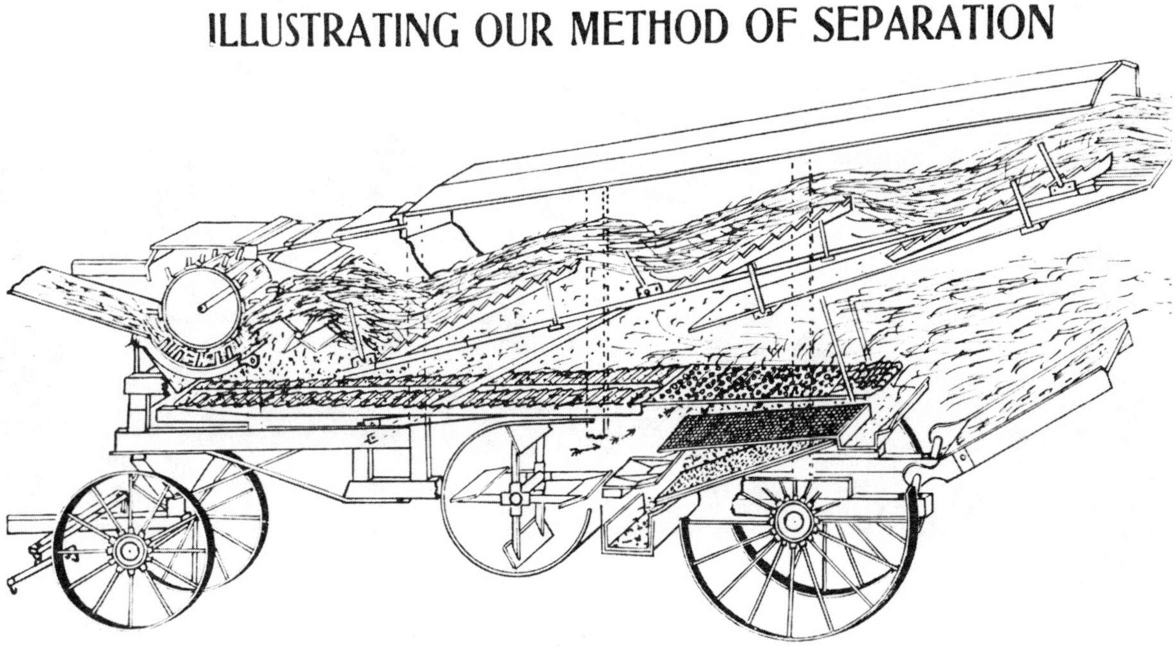


*This cutaway view of an early twentieth century threshing machine shows how the implement worked. The crop was fed into the machine at the left. The spiked concave threshed the grain from the heads, and the grain fell below onto a winnowing or separating rack. The straw was carried out of the machine on the top racks. The fan at the bottom blew the chaff away, and the cleaned grain poured out of the thresher at the bottom.*

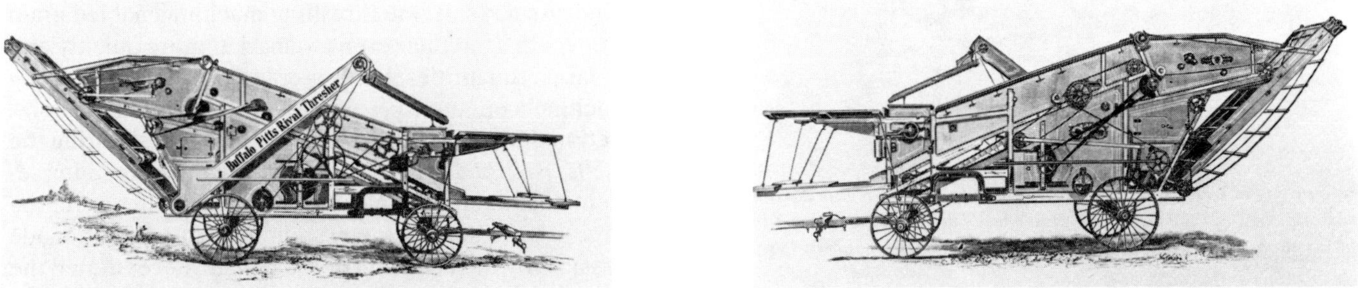


*In 1894, the Buffalo Pitts Agricultural Works placed this threshing machine on the market. The stacker elevated at the rear to carry the straw away. This machine was designed for steam power.*

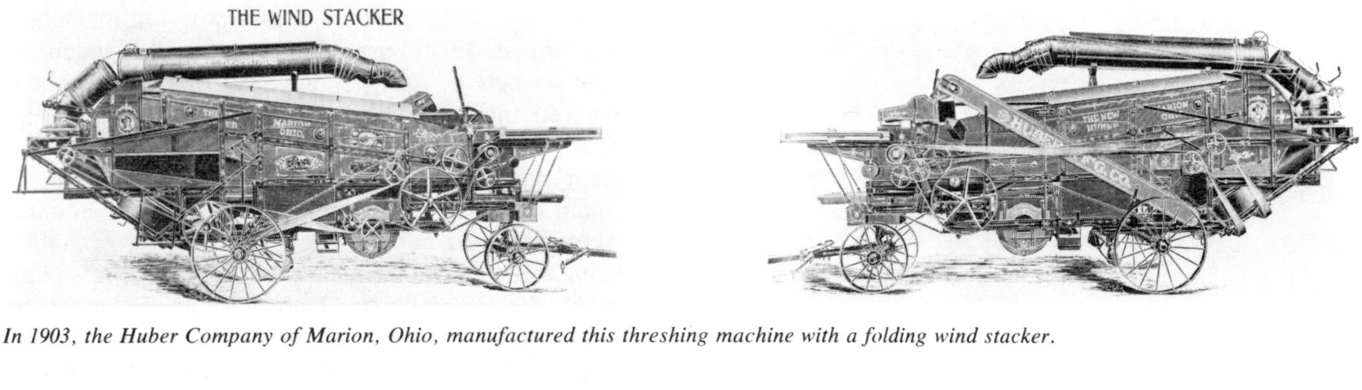


*In 1903, the Huber Company of Marion, Ohio, manufactured this threshing machine with a folding wind stacker.*

*This threshing scene shows a wind stacker in operation. (Smithsonian Institution.)*

THREE-HITCH GEAR, PIVOT CYLINDER, DOUBLE FAN and STACKER

*Soon after the Civil War, implement manufacturers began adding straw stackers to their threshing machines. The Russell & Company of Massillon, Ohio, offered this thresher and stacker for sale in 1869.*

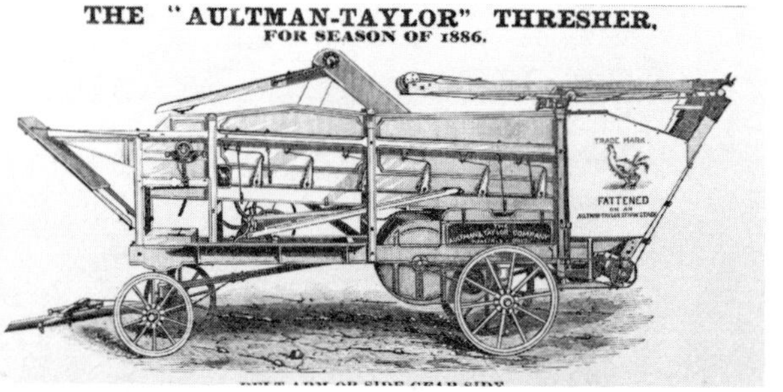


*In 1886, the Aultman-Taylor Company placed this threshing machine on the market. It was suited for either horse or steam power.*

could be completed during the winter when there was an abundance of cheap farm labor available. By the 1850s, however, these arguments had largely disappeared. Threshing machines had, by that time, proven themselves more efficient than the flail, because the machine did the work faster and threshed more thoroughly than when the work was completed by hand. As a result, the threshing machine enabled grain farmers to send their crops to market more quickly and in larger quantities than ever before. The threshing machine's popularity was also enhanced with the completion of the railway system in the Midwest during the 1850s. Railroads gave farmers greater access to implement manufacturers than previously. Furthermore, the large wheat acreages west of the Ohio River, made possible by the adoption of the reaper, necessitated the adoption of more efficient threshing methods. Consequently, by the mid-nineteenth century, implement companies were producing thousands of threshing machines annually with the Pitts, Emery, Westinghouse and Case machines dominating the field.

During the early 1870s, steam engines increasingly powered threshing machines. In less than a decade, steam had almost entirely replaced horses for power. By the turn of the century, steam-powered machines could thresh 3,000 bushels of wheat and as much as 6,000 bushels of oats per day. By that time, horse-powered threshing machines were no longer being used. Although the gasoline tractor would replace the steam engine in the 1920s, the threshing machine would remain a necessary implement for the grain farmer until tractor-powered combines became readily available during the 1930s. In the meantime, the giant straw pile, left by the threshing machine, became a symbol of prosperity, protected cattle against the winter winds, and provided a forbidden play-mound for farm children.

*Between 1836 and 1854, Hiram Moore built five combines. This picture, made from one half of a stereopticon slide, shows Moore in a black top hat sitting on one of his combines. Ca. 1850-1852. (Smithsonian Institution.)*

# CHAPTER VIII

# THE COMBINES

Not long after Cyrus Hall McCormick and Obed Hussey patented their reapers, and about the same time the Pitts brothers were experimenting with their thresher-cleaner, Hiram Moore built the first successful combine. The combine is simply a combination of a reaper or harvester with a threshing machine and winnower attached. The first combine patent, however, was not granted to Moore. In fact, three patents were given to other inventors before Moore, together with his financial backer — John Hascall, received a patent on 28 June 1836. On 8 August 1828 Samuel Lane of Hallowell, Maine, became the first person to patent a combined harvester, but he probably never built that machine for field testing or sale. Similarly in 1835, a D. Ashmore and a J. Peck of Tennessee patented another combine which evidently did not get much farther than the drawing board. Apparently this combine looked much like the Gallic reaper, but little more is known about it since the patent records have been destroyed. Several weeks before Moore received his patent, E. Briggs and C. G. Carpenter actually built their combine and tested it in the vicinity of Rochester, New York, in 1839. Four large wheels supported this machine, the back two of which powered the gearing for the threshing cylinder and winnower. With two yoke of oxen and a team of horses for draft power, the Briggs and Carpenter combine could harvest about twenty acres per day. Still, this combine did not function well enough to merit widespread attention, and the combine's feasibility did not become apparent until Hiram Moore began field testing his machine.

In 1831, Moore moved from Vermont to Climax Prairie in Kalamazoo County, Michigan, and began farming. Sometime soon thereafter, John Hascall, a neighbor, urged Moore to design a combine. Hascall's wife had had a dream in which she saw a giant horse-drawn machine cutting and threshing grain all at once. Hascall knew that Moore had an inventive mind and told him about the dream. Although it is uncertain when Moore began working on the problem of designing a combine harvester, he completed a patent model in 1834, and field tested his combine near Flowerfield, Michigan, the next year. At that time, however, Moore's combine was incomplete since the threshing cylinder was not attached, and the machine cut only two rods before it broke. Moore returned to his shop to revise his plans and to make repairs. In October of that same year (1835), he was back in the field for another test. This time he had the threshing cylinder attached but not the winnower. Moore's combine worked to the delight of the bystanders. Twelve horses pulled the machine which cut and threshed three acres of grain at an estimated cost of $.82 per acre — a considerable savings over the going rate of $3.12 for cradling, raking, binding, shocking, threshing and cleaning grain.

Moore realized this modest success did not mean he had perfected the combine harvester. Improvements and additional tests were needed. In July 1838, for example, Moore's twenty-horse, fifteen-foot swath combine cut thirty acres near Climax, Michigan. A year later, Moore was back in the field at Prairie Rhonde where his combine cut sixty-three acres at the rate of twenty acres per day. Although Moore's combine worked, the short Michigan summers did not provide a

*Combine advertisement from the **Pacific Rural Press**, April 27, 1889. (Smithsonian Institution.)*

harvest time of sufficient length for extensive testing. Nevertheless, by 1843, Moore believed he had built a combine that was a practical harvesting machine. At that time, his combine could harvest twenty-five acres per day.

Between 1836 and 1854, Moore built five combines with the mechanical help of the tool-and-die makers in Rochester, New York, and in Schoolcraft and Battle Creek, Michigan, and with the financial aid of Hascall and several other investors. Those combines looked far different from the self-propelled combines of today as well as from the mammoth horse-drawn combines which inventors developed in California late in the nineteenth century. Essentially, Moore's combine was a huge threshing machine, mounted on wheels, with a reciprocating sickle and a gathering reel. This combine was seventeen feet long and fifteen feet wide. Two wheels, seven feet in diameter, with iron spikes to prevent slipping on wet ground, provided the power for the cutting and threshing mechanisms. The cutter bar consisted of a fixed plate with saw teeth and a saw-tooth sickle which oscillated on top. The cutter adjusted to permit shearing the heads close to the top of the stalk, since the height of the grain varied from field to field or even within the same acreage. A gathering cylinder measured four feet in diameter and twelve feet in

*Wheat farmers needed at least a twenty-horse or mule hitch to pull the combines made during the late nineteenth century. The Holt Brothers of Stockton, California, made this combine. In 1908 it was at work in a California wheat field. (National Archives.)*

length. Rows of six-inch wooden or metal teeth extended from the cylinder. These teeth caught the grain and pulled it into the reciprocating sickle. An apron carried the cut grain to the threshing cylinder. Behind the cylinder a revolving wire riddle separated the straw and dropped it onto the ground. The threshed grain and chaff fell through a sieve where a winnowing fan blew away the chaff. The grain then passed through a conveyor and spout into bags. A dividing bar on the edge of the machine separated the standing grain from that which was about to be cut. The sickle and reel were adjustable to permit cutting grain of various heights. At least six men were required to operate Moore's combine. One rider was also needed for each four-horse team. Several additional workers were needed to haul the grain sacks to the barn or granary.

By the late 1840s, Moore had changed the combine's design so that a huge, barrel-like threshing cylinder was set at an angle behind the reel. The separated straw dropped to the ground from the cylinder's open end. The grain fell through a screen at the bottom where a fan winnowed the chaff. The cleaned grain collected in a bin from which it was elevated to a platform where a worker, riding on the machine, collected it in bags. This machine, pulled by sixteen horses walking two abreast, cut a ten-foot swath, and sacked the grain at the rate of twenty-five acres per day. Moore's combine worked best in fields free of stumps and large rocks and was reportedly simple enough for operation by "any man of ordinary common sense after two day's experience."

Moore continued to make adjustments on his combine, but the midwestern farmer did not adopt it nor was it manufactured by any agricultural implement company. Four major problems prevented the acceptance of Moore's combine in the midwestern grain belt. First, the fields were small, and Moore's combine was large, clumsy, and difficult to maneuver. The lighter, simpler, more manageable reapers just coming on the market required only two horses for draft instead of sixteen or twenty, and they were far more suitable for the grain farmers' needs. Second, the humid climate and wet summer weather of the Midwest kept the grain stalks from drying. Damp stalks and tough grain heads did not thresh properly and often clogged the cylinder. However, if the grain was cut with a reaper and placed in shocks it would cure and dry out enough for machine threshing. Third, rain could delay the harvest for days or weeks if the ground was too muddy for the combine to operate properly. While a farmer waited to get his combine into the field, more severe weather might destroy the entire crop. Fourth, Moore estimated that his combine would cost $500 each, far more than the Midwestern grain farmer could afford.

Midwestern farmers weighed the disadvantages of Moore's combine against the advantage of quick completion of the harvest by cutting, threshing and winnowing all in one operation. Virtually all of them decided the risks of breakdowns and bad weather and the expense of a machine which required far more draft horses than most farmers owned were not economically feasible. Eventually, midwestern farmers would adopt the combine, but they would not begin to do so for another half century. At the turn of the twentieth century, small combines which were technologically suitable for midwestern fields and crops would reach the market. In the meantime, the reaper and the thresher satisfied their harvesting needs.

In 1853, John M. Horner, a bonanza wheat farmer near Mission San Jose, California, invited A.Y. Moore and George W. Leland, who had purchased one of Hiram Moore's combines, to send it to the West Coast and test it in the California wheat fields. Moore and Leland accepted the invitation and sent the combine to California via New York and Cape Horn. Oliver Kidwell Moore, A. Y. Moore's son, met the combine at San Francisco and used it to cut 600 acres of wheat in Alemeda County during the 1854 harvest. Moore's combine attracted a great deal of attention. An editorial in the *California Farmer* reflected with pride that: "This is one of the most wonderful inventions of the age and the sight of it is well worthy of a visit to this great valley." Moore's combine was not used during the 1855 harvest, and it burned in the field the following year when a bearing overheated after a negligent operator failed to lubricate the machine.

In spite of that disaster, Moore's combine had proven well suited for agriculture on the West Coast, particularly in the San Joaquin Valley where the dry harvest season and large wheat fields made combining practical. There, the combine's success created a flurry of further inventions to improve the machine. Actually, James E. Patterson, a California inventor, constructed a combine in 1852, the year before Moore's machine was shipped west. This combine required twenty-two mules for draft power. When the combine was first tested, however, the clatter of the machine caused the mules to bolt. Before the runaway team could be halted, the combine had been torn apart, and it was never reassembled.

Between 1853 and 1866, John M. Horner, a financial backer of Patterson's, also built three combines. Horner tried to eliminate the potential problem of a runaway team destroying the combine by harnessing the horses or mules behind it in a manner similar to that used on the header. With the Horner "Traveling Harvester Monitor No. 2," three men and twelve horses could cut, thresh, clean, and sack fifteen acres of wheat per day at half the cost of binding and threshing. The *California Farmer* was so impressed with this combine's performance, during an 1868 field test, that it urged farmers to see it for themselves even if they had to travel several hundred miles to do so. Some farmers believed combines would have revolutionary effects on farm labor costs. Many farm workers evidently believed this as well because an arsonist, presumably a harvest worker, burned one of Horner's combines in the field the following year.

In 1876, David Young and John C. Holt of Stockton,

*By the 1880s, combines had been fitted with adjustable wheels and levelers, thereby, enabling the machines to harvest rolling or hilly lands. This hillside combine was at work in the state of Washington in 1888. (Smithsonian Institution.)*

*Some horse-powered combines were used well into the twentieth century. In 1925, twenty-six head of horses and mules pulled this combine across a vast wheat field in Umatilla County, Oregon. (National Archives.)*

California, built a similar combine which they called the Centennial Harvester. Twenty-four horses or mules, hitched twelve abreast, pushed the machine through the wheat fields. Two wheels generated the power to turn the combine's gearing. The left wheel drove the thresher and separator, while the right wheel powered the sickle bar and reel. Gears powered, however, by ground wheels wore out quickly or broke frequently. This problem, though, was solved by the Holt Manufacturing Company between 1885 and 1889, when agricultural engineers began using link chains and V-belts instead of gears to transfer power from the wheels to the various turning parts. If a chain or belt broke, it could be repaired relatively easily.

Other inventors tried their ability at designing and building a workable combine and more than twenty types of combines were made in California between 1858 and 1888. Most of those inventors got no farther than the patent office. Although one optimist, writing in 1878, believed that flour eventually would be milled on the combine, few machines were in use by 1880. Indeed, it was not until 1883 with the organization of the Shippee Harvester Works that the business of making combines began to pass from the workshops of individuals into the machine shops of the major agricultural implement companies. L. U. Shippee, who directed the company, purchased combine patents from many inventors and proceeded to build a machine that would meet the needs of the average California farmer at a reasonable cost. In 1884, the Haines and Houser Company merged with the Shippee firm to form the Stockton Combined Harvester and Agricultural Works. This new company manufactured not just one combine but a number of machines based upon the patents which Shippee had purchased earlier. One of the most popular combines produced at the Stockton factory was the Houser. By 1886, 280 Housers had been produced which harvested more than 300,000 acres of wheat that year. These machines cut and threshed from 25 to 35 acres per day and cost from $1800 to $2000 a piece. The combines produced at the Stockton Combined Harvester and Agricultural Works, such as the Shippee, Powell, Minges, Gratten, and Benton, had standardized parts. The development of standardized combine parts meant that if a part broke in the field, it could be replaced relatively quickly, since it did not have to be specially made to meet the specifications of a particular machine which was itself a unique implement. Standardized or interchangeable parts meant that less time was lost during breakdowns. Besides the machines produced at the Stockton Combined Harvester and Agricultural Works, the Young, Reynolds and Patterson, Matteson and Williamson, Marsters, Meyers, Holt, Herald, Price, and Judson were popular along the West Coast, during the last quarter of the nineteenth century. The Judson combine was the unique machine of this group. It was powered by both a steam engine and horses. Ten horses pulled the combine while a 12-horsepower oil-burning steam engine drove the reel, cutter, thresher, separator and bagger. This combine was probably the first farm implement to use oil for fuel.

As early as 1871, B. F. Cook, another California inventor, attached a steam engine to a Haines header and a Pitts thresher. While the steam engine powered the machine, a team of horses pulled the combine through the field. Although this machine did not work particularly well, it introduced the age of steam to combine harvesting The first self-propelled steam combine, however, did not appear until 1886. That innovation, conceived by George S. Berry, consisted of a 26-horsepower Mitchell-Fisher steam engine (operating in a reversed position) providing forward drive, and a 6-horsepower Westinghouse steam engine driving the cutting, threshing, and winnowing mechanisms. The engine mounted on the combine received its steam

*Combining with a twenty-five horse hitch near Walla Walla, Washington, 1917. (National Archives.)*

through a flexible tube attached to the traction engine's boiler. Straw from the thresher fed back in a chute to the boiler firebox, thus making this "straw burner" steam engine the first of its kind. At first, the sickle on this combine cut a twenty-two foot swath, but by 1888, Berry had extended it to forty feet. This combine cut fifty acres per day. The next year (1889), Daniel Best's Steam Harvester also appeared on the market. Steam from the tractor's boiler powered an auxiliary engine of the combine, and in 1891, Benjamin Holt also marketed a similar steam-powered combine. For the next twenty years combines were manufactured with auxiliary steam engines which took steam through a flexible tube from the traction engine which pulled the machine. The auxiliary engine, using the steam from the lead tractor, powered the cutting, threshing, separating, winnowing and bagging mechanisms. In 1904, the Holt Company began substituting gasoline engines for steam engines and, by 1912, the internal combustion engine had replaced both the auxiliary and the traction engine for most combine work.

Steam-powered combines required a crew of six to seven men. The driver, fireman, and water hauler attended the steam traction engine. The header controlled the sickle's height, and the tender made sure the cut grain fed into the threshing cylinder evenly to prevent clogging. A sacker filled the bags, sewed them tight, and dropped them onto the ground. Finally, a man and a team might follow the combine and load the sacks onto the wagon for transport to a railway loading dock or to storage facilities elsewhere. Some of these giant machines, weighing fifteen tons with forty-foot sickle bars, cut 100 acres and threshed 2,500 bushels of wheat a day. Even with harvesting capacities such as these, steam-powered combines were less efficient than horse-drawn models, because the extra men required to man the steam engine increased harvesting costs far beyond the expense of the three men needed to operate the horse-drawn combines.

In addition, steam-powered combines were dangerous, because a spark from the fire-box or smokestack could set the entire wheat field ablaze. Horse-drawn combines could perform nearly the same amount of work as those pulled by steam engines. A forty-horse hitch with a thirty-foot cutter bar could harvest from seventy-five to one hundred acres per day with no danger from sparks. Two ground wheels powered the horse-drawn combines. One wheel usually drove the sickle bar and the other powered the threshing and cleaning mechanisms.

No matter, though, whether these large combines were horse- or steam-powered, they were suitable only for flat lands. Wheat fields planted on hilly ground still had to be cut with reapers, headers or binders, because the gigantic combines toppled over when used for hillside harvesting or else the long cutter bars dug into uneven ground. By the 1880s, however, inventors had added adjustable wheels and levelers which raised and lowered the combine's sides as conditions dictated in order to keep the machine balanced. Most of these hillside combines were designed for horse-power because only track-type tractors could safely operate on steeply sloping ground without danger of tipping over.

During the 1890s, combines, whether steam- or horse-powered almost completely replaced the header on the bonanza wheat farms in California. By the turn of the twentieth century, the combine harvester cut approximately two-thirds of California's wheat crop at a cost of approximately $1.75 per acre. Only a decade earlier, harvesting costs, when using a header and threshing machine averaged $3.00 per acre. By the beginning of World War I, more than forty combine makes were produced from assembly lines in California, Idaho, Oregon, and Washington.

Certainly, the combine increased productivity and reduced labor costs, because it wasted less grain than reapers and threshers and fewer hired hands were needed at harvest time. Indeed, for the farmers who could afford the machine, the combine made them almost independent from hired help. Not only was the crop removed at once and the field thereby cleared for immediate plowing, but the cut straw was also scattered across the ground to help build soil humus. Furthermore, the combine freed farm women from the drudgery of cooking meals for large threshing crews.

By the turn of the twentieth century, the combine was still primarily used only along the West Coast. Most of the combines used east of the Rocky Mountains prior to World War I operated in the Great Plains states where the dry climate made the wheat suitable for combining. Wartime labor shortages, though encouraged farmers to buy combines. During the 1920s, technical advancement, which improved maneuverability and reduced the combine's size, and the adoption of the gasoline tractor, which gave the combine consistent drive with its power take-off, helped make the combine popular in the Midwest by the late 1930s. Until that time, however, combine harvesting remained a mechanical phenomenon of the West Coast.

# CHAPTER IX

# MAKING HAY AND FODDER

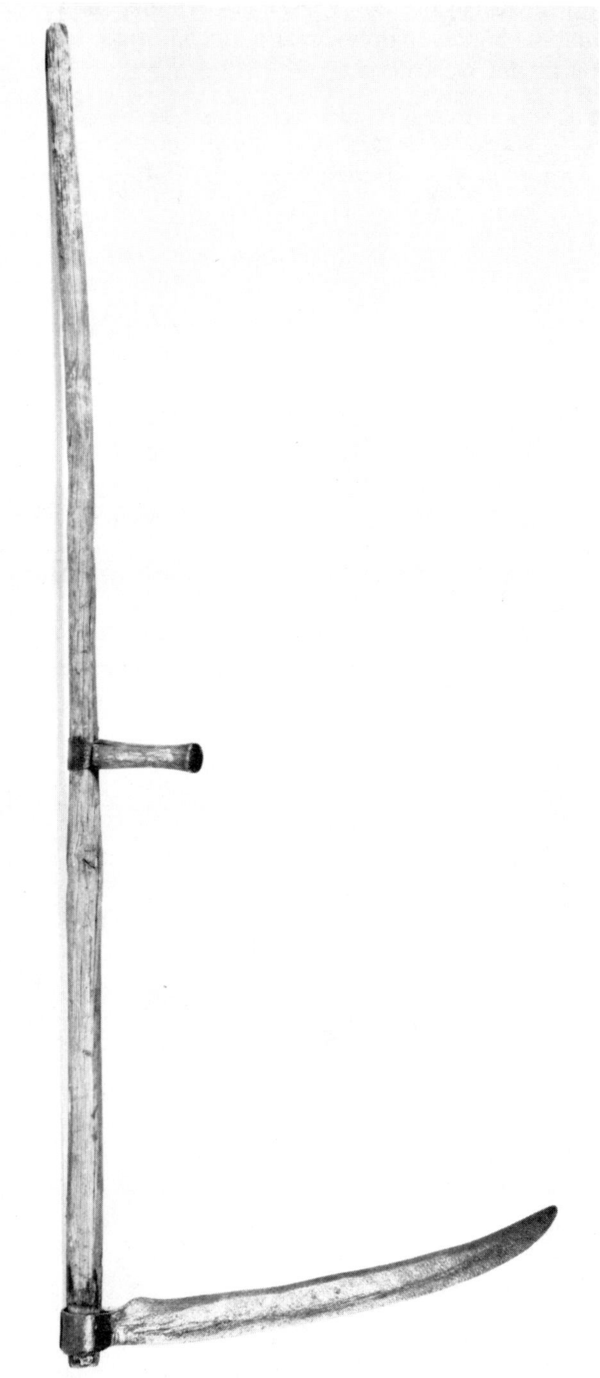

*Until the introduction of the horse-drawn mowing machine in the 1830s, farmers used the scythe to cut hay. (Smithsonian Institution.)*

Prior to the age of the internal combustion engine, hay was the most widely grown crop on farms devoted to general agriculture. Farm, city, and military horses and mules required large amounts of feed and the annual hay crop was a primary cash source for farmers who owned grasslands and meadows. The economic value of the hay crop caused anticipation, worry, and disappointment, depending on the outcome of the weather at haying time. In addition, the job of making hay became even more worrisome, because the haying season came during the harvest time for the small grain and corn crops. If the grass was cut but not raked and stacked because of labor shortages, or because attention was given to other pressing demands, it might be ruined by bad weather. A drenching rain would cause hay to mold if it were not properly cared for, or some of the nutritional value might be lost — all of which meant a substantial monetary loss for the farmer. Furthermore, since two and sometimes three crops of hay could be cut each season, the problems of hay-making were compounded.

## HAYRAKES

Before the nineteenth century began, a man could cut one or two acres of hay per day with a scythe, but once the hay was cut it still had to be raked, loaded onto a wagon and hauled to the barn. Making hay in this manner required little equipment; a scythe, wooden rake, pitchfork, and a wagon were all the tools needed. Although the investment in hay-making implements was minimal, the job of haying was hard, slow work.

Although the scythe was universally used on farms until the introduction of horse-drawn mowing machines in the mid-1830s, technical change came early to the hay-making process. Sometime before the turn of the nineteenth century, the exact time of which is uncertain, Virginia farmers began using horse-drawn rakes which they either imported or copied from the British. These horse-rakes had large wooden combs about ten feet wide with teeth two feet long. The teeth were spaced approximately eight inches apart and plow handles were attached to the frame. As the horse drew the rake across the field, the farmer manipulated the rake over rocks and around stumps with the handles. The grass gathered on top of the teeth which ran flat on the ground. When the rake was full, the horse was reined to a halt, the rake lifted and the hay shaken loose in a heap or windrow. Once the hay was in windrows, it could be pitched onto a wagon and hauled to the barn. Sometimes the horse-rake was drawn along the windrow, and the hay collected and deposited at the place in the field where the stack was to be built. These hay rakes could replace from three to six hired-hands in the field.

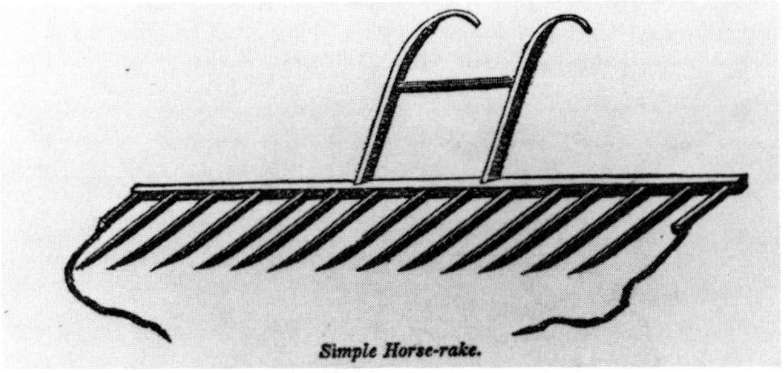


*Prior to the nineteenth century, farmers used horse-rakes, such as this one, to put hay in windrows. The rope on each side of the rake attached to the harness.*

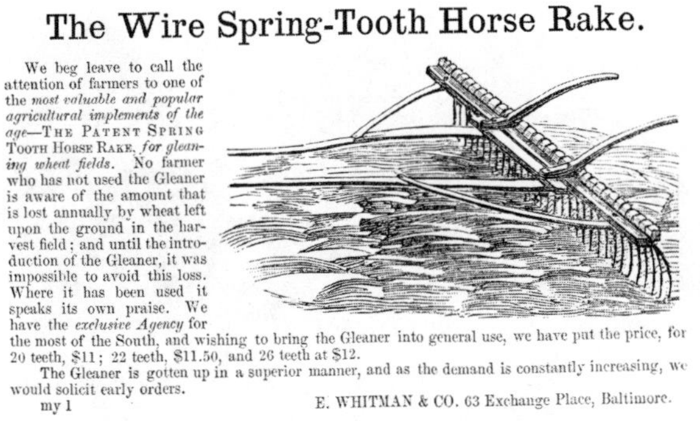


*Advertisement for a spring-tooth rake. (Smithsonian Institution.)*

*American farmers began using the revolving horse-rake during the early nineteenth century. When the front teeth filled with hay, the farmer lifted the handle and the back teeth rolled over to gather a new swath.*

By 1820, farmers in the middle Atlantic states and New England had adopted the horse-rake, but it was seldom used in the South where farmers preferred to graze their cattle year round and to use cottonseed cake for supplementary feed. The wooden horse-drawn hay rake became extremely popular, however, in New York, Pennsylvania, and New Jersey where the terrain was well suited for this implement. The rocky New England soil was less satisfactory, because the horse-rake was difficult to maneuver over it and the wooden teeth broke frequently; consequently, it had only limited appeal in that region.

Sometime before 1820, the revolving horse-rake appeared. Although the exact origin of this rake is unclear, Ephraim Perkins and Charles Gouge of Oneida County, New York, made such an implement in July 1811; and, by 1823, Samuel Pennock at Kennett Square, Pennsylvania, was manufacturing rakes of this type for commercial sale. The revolving hay rake looked like the simple horse-rake, except that it had a second row of teeth mounted on the beam directly behind the teeth set to gather the grass. When the first set of teeth filled with hay, the farmer lifted the handles which caused the beam to revolve. The teeth loaded with hay kicked under and back and the entire rake revolved over the hay which was left in a pile. At that same time, the second set of teeth immediately revolved up and over and began to gather more hay. The revolving hay rake eliminated the need to stop the horse and lift the rake over the windrow to empty the cut grass or clover from it.

Farmers slowly adopted the revolving horse-rake. By 1840, however, they were using it in every state north of Virginia; and, by 1850, it had become a standard implement wherever hay was made. With it a man could rake a ten-foot swath into windrows as fast as he could walk, and cover about three acres an hour. Simply put, the revolving horse-rake replaced about six men with hand rakes.

About 1850, implement manufacturers began adding seats to their rakes. Although the first sulky rake appeared at least as early as 1837, it was not until 1849, when Calvin Delano of Maine patented his sulky rake, that this implement became popular. Delano's sulky rake had wooden teeth which the operator raised or lowered with a lever. By the mid-1860s, "Warner's Sulky Revolver" was one of the most popular rakes. This rake was manufactured by the Blymer, Day & Company of Mansfield, Ohio. A lever, extending from the rake to the driver's seat, enabled the operator to trip the load when the rake was full. The teeth were tipped with iron to prolong wear. When the job was completed, the rake could be detached and fixed in front of the driver to facilitate travel down country roads.

During the 1860s, sulky rakes with spring-teeth also became popular. The Hollingsworth, made by the Wanzer & Cromwell Company in Chicago, was one of the most popular sulky rakes of this type. Rakes with steel teeth could be used on rocky ground with less damage than rakes with wooden teeth. After 1870, the

The wooden-tooth sulky rake became popular during the 1850s. By the next decade, however, most sulky rakes were fitted with spring-teeth.

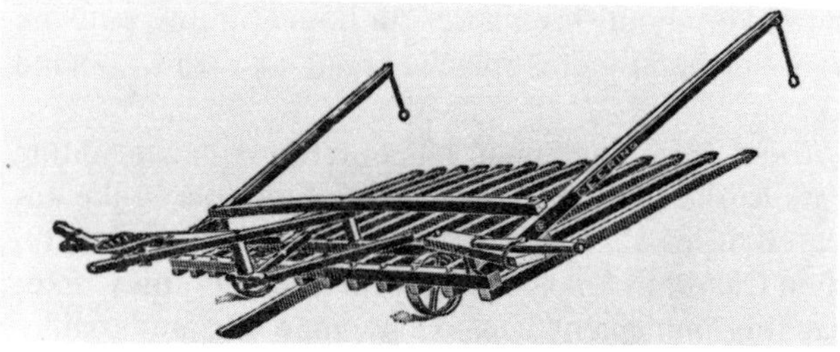

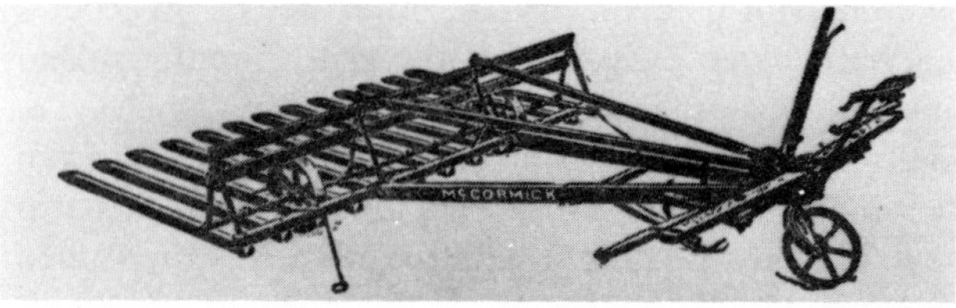

During the late nineteenth and early twentieth centuries, farmers with large hay fields used two-wheeled sweep-rakes to gather hay. The sweep-rake collected about a half ton of hay before it was full.

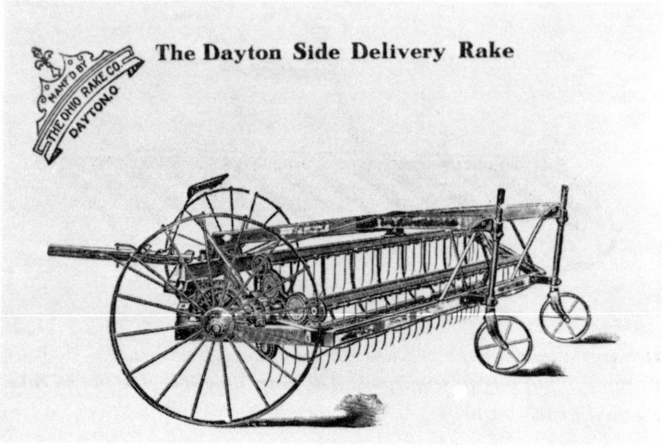


During the late nineteenth century, farmers increasingly adopted the side-delivery rake, because it did the work faster and more efficiently than any other rake.

spring-tooth sulky rake remained virtually unchanged. With ten- or fifteen-foot sulky rakes, a farmer could rake twenty to thirty acres of hay per day. The sulky rakes were often known as hand-dump or self-dump models. The farmer tripped a lever on the hand-dump rake which emptied it in the windrow. The self-dump rake emptied when a foot pedal was tripped. The operator had to watch the dump rake constantly in order to release the hay at the proper time. Even so, the windrows left by the dump rake often zigzagged across the field. Crooked rows meant the crew loading the hay with pitchforks had more work to do as they walked back and forth between the windrows and the wagon. Still, the sulky rake eliminated the need for the operator to walk behind the horse-rake and guide it with the handles. By so doing, sulky rakes greatly speeded the haying process. The most popular sulky rakes were those made in eight- or ten-foot widths.

The side-delivery rake, which appeared after the Civil War, was an even more useful implement than the sulky rake. The side-delivery rake came in two styles. One consisted of a wheeled frame with spring teeth mounted on a reel. The reel was set at a forty-five degree angle and as the wheels turned the gearing, the teeth kicked the hay into a windrow off to the side of the rake. The second style had wheel rakes set at an angle which pushed the hay off to the side and into a continuous pile. By the First World War most farmers preferred the side-delivery to the dump-rake, because it worked faster and more efficiently. The side-delivery rake did not work well on rocky ground, but it became increasingly popular in the Midwest after 1890. There, the level prairie lands and heavy grasses made the side-delivery rake a welcome farm implement.

By the late nineteenth century, the sweep-rake, sometimes called a buck-rake, bull-rake, go-devil or slip-around, came into extensive use in the Middle West and Great Plains. This rake looked much like a revolv-

*In February, 1822, Jeremiah Bailey patented this mowing machine. The circular blade did not cut grass efficiently, and it never became a pratical mower. (Smithsonian Institution.)*

ing horse-rake. It had teeth five feet long and a wooden box-like frame on top. A horse was attached to each side, and, as the rake was pulled down the windrow, the frame filled with hay. Once full, the rake was pulled to the stack or the barn and the horses turned about. This caused the hinged sides to open and reverse position. The rake was then pulled away ready to gather another load. This rake reportedly collected thirty tons of hay in a ten-hour day.

At the turn of the twentieth century, the sweep-rake consisted of a horizontal beam, usually about twelve feet long, fitted with long wooden teeth. The teeth were slightly elevated and had steel caps to prevent them from digging into the ground as the rake was pushed ahead of a team of horses. The teeth were eight feet long and were spaced one foot apart on the beam. The teeth slipped under the hay and collected it against a brace at the rear. The sweep-rake came with or without wheels. Wheelless and two-wheeled sweep-rakes were pulled with a horse hitched to each side. Sweep rakes with three or more wheels were pushed ahead of the horses. The wheelless sweep-rake was made for hauling hay to the stack in the field, since it was unsuitable for travel along country roads. Wheeled sweep-rakes were designed so that the teeth could be raised after the hay had been swept up or loaded. This placed the weight directly on the wheels and eased the burden on the horses. The wheeled sweep-rake also kept the hay cleaner and required less draft power, during transport to the stack or barn, since it did not drag across the ground.

Although the sweep-rake was occasionally used to gather hay directly from the swath, it was most commonly used to gather the grass or clover after it had been raked into a windrow. The sweep-rake reduced the cost of hauling hay from the windrow to the stack or barn by fifty percent, and one boy could operate it alone. A two-man crew using sweep-rakes could haul twice as much hay to the barn as a three-man crew using pitchforks and a wagon, or three times as much hay if it was stacked in the field. The sweep-rake required heavy, powerful horses to operate efficiently, because the rake might collect a half ton of hay before it was filled and ready for dumping.

## MOWING MACHINES

Until the early nineteenth century, the American farmer cut his hay with a scythe in a manner little changed since antiquity. On 4 December 1812, however, Peter Gaillard of Lancaster, Pennsylvania, patented a horse-drawn machine specifically designed for mowing grass. By doing so, Gaillard began a new age for haying technology, even though his mower proved unsuccessful. Ten years later, on 13 February 1822 Jeremiah Bailey of Chester County, Pennsylvania, became the second American inventor to patent a mowing machine. Bailey's mower consisted of a rotary blade five and a half feet wide which looked much like a modern circular saw blade. The blade was attached to a

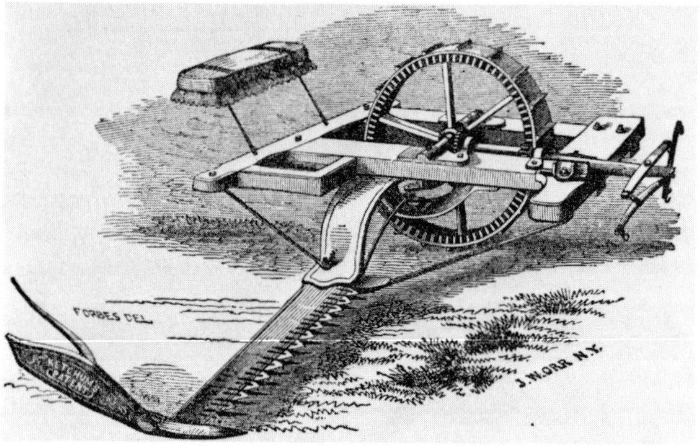


*Left: The Ketchum mower became the most popular single-wheeled mowing machine during the nineteenth century. (Smithsonian Institution.)*

*Below: Mowing machine advertisement from* **The American Farmer-Advertizer**, *1855. (Smithsonian Institution.)*

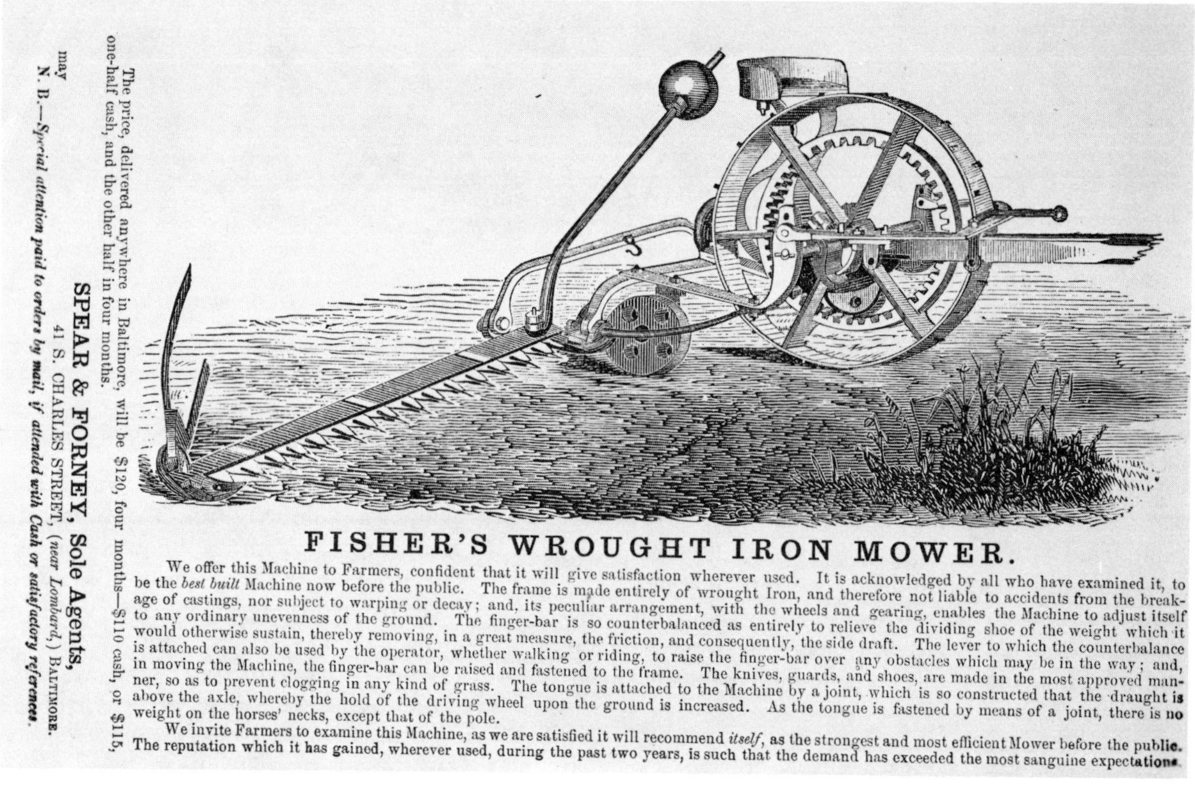

*In 1884, the McCormick Harvesting Machine Company offered this mower for sale. Note that the gears at the center of the axle have been covered to keep out the dust.*

*The flexible cutter bar was a standard feature on mowing machines after the mid-1850s. In 1886, this mower was manufactured by J.F. Seiberling & Company of Akron, Ohio.*

spindle which was connected to gearing on the drive wheel. A weighted lever enabled the farmer to adjust the height of the blade. The gearing, blade and lever were mounted on a two-wheeled frame made from white oak. Bailey claimed his mower would cut six acres per day and replace twelve men with scythes in heavy grass and six men in light grass. Although Bailey's mower attracted some attention in Pennsylvania and New Jersey as well as in England, it too did not function efficiently enough to merit production and widespread use. Indeed, the horse-drawn mowing machine did not become practical for more than twenty years after the first patent had been granted on that implement.

On 3 May 1831 William Manning of Plainfield, New Jersey, patented a mowing machine which proved successful. Manning's mower incorporated for the first time the principle of the reciprocating sickle. Manning's sickle or cutter rested on the ground, and as the machine moved forward, the sickle reciprocated as it moved through the grass. Iron teeth, six or eight inches long, protruded from under the sickle and gathered and held the grass for the oncoming blade. Manning's mower, like Gaillard's and Bailey's before him, failed to work sufficiently to merit adoption by a large number of farmers. Nevertheless, Manning's reciprocating sickle indicated great possibilities to other inventors, who believed that, with certain adjustments, it could be made to work efficiently.

Obed Hussey was the first inventor to improve upon Manning's concept. Hussey's combination reaper-mower, patented in 1833, had a superior sickle for cutting grass; and after some adjustment in 1847, it became the standard cutter for both reapers and mowers. Combination reaper-mowers, however, did not work entirely satisfactorily. The reaper platform either interfered with mowing, or, if it was removed, the platform's absence weakened the machine. Because of these problems, inventors began designing implements specifically intended for mowing hay.

On 10 July 1847 William F. Ketchum of Buffalo, New York, patented a durable mower designed specifically for cutting hay. Ketchum's mower was simply built with a single drive wheel and a cutter bar which consisted of an endless chain of knives. A small runner supported the outer edge of the sickle. The endless cutting apparatus did not prove successful, but after Ketchum adopted Hussey's sickle, his machine became the first mower manufactured in large quantity. In time, the Ketchum mower earned the reputation of being the best single-wheeled mowing machine ever produced.

Beyond the addition of a second wheel, the two-wheeled mowing machine had a flexible cutter bar instead of the fixed sickle found on the single-wheeled mowers. The origin of the flexible cutterbar can be traced to Hussey's reaper-mower which had a hinged sickle attached to the main axle. On 5 December 1854 Cyrenus Wheeler patented a two-wheeled mower with a flexible sickle design which proved a lasting success. Wheeler placed the cutter bar at the rear of the machine. This location permitted the driver to raise or lower the blade with a lever while the mower was in motion. The D. M. Osborne & Company of Auburn, New York, manufactured Wheeler's mower and marketed it under the trade name of "Cayuga Chief." The next major contribution in the development of the flexible cutter bar came on 4 September 1855 when Jonathan Haines patented a mowing machine with two drive wheels.

*Flexible cutter bars enabled the operator to work around obstructions without loss of time. In 1882, this mower was built by the D.M. Osborne & Company of Auburn, New York.*

The mowing machine's design was further improved on 17 July 1854 when Cornelius Aultman and Lewis Miller of Akron, Ohio, patented a two-wheeled mower. In May 1853 Miller patented another design with a floating cutter bar, that is, a sickle which followed the contours of the ground and thereby cut more efficiently. Since that time, the design of the mowing machine has changed little. Aultman and Miller marketed their mower under the name "Buckeye." The Buckeye mower differed from previous machines in several respects. First, a series of gears, attached to the main axle, drove the sickle. This feature gave better balance to the machine since gearing fixed to only one drive wheel caused the machine to veer off to one side. Clutches on each drive wheel enabled the operator to engage or disengage the wheels at the proper moment and thereby turn sharper corners than ever before. The cutter bar was mounted in front of the driver. This was an important safety feature, since the sickle was fixed behind or off to the side of the driver on the other machines. If the mower hit a hole or obstruction, the driver could be easily thrown into the path of the sickle. By placing the blade in front of the driver, the operator was out of the path of the oscillating sickle if he fell from the machine. The cutter bar could be raised or lowered with a lever. The Buckeye mower had an additional convenience because the sickle raised and folded in front of the driver. This feature facilitated traveling down country roads and passing through narrow gates. The Buckeye mower was an immediate success; and, as long as hay was mowed by horse-power, it remained one of the most popular implements for cutting grass and clover.

The mowing machine improvements which followed the Buckeye mower were primarily designed to lighten, strengthen, and reduce the cost of the machine. By 1860, the mower was a practical farm implement, and the Cayuga Chief, the Ball, and the Buckeye were the most popular mowing machines at that time. These mowers cut a narrow swath, and, as late as 1918, the most commonly used mowing machines had only five-foot cutter bars. These mowing machines cut an average of ten acres per day, and thereby replaced ten men with scythes. The two horses required to pull the mower were also cheaper to feed than a mowing crew. One farmer believed the mowing machine made the job of cutting hay so easy that it could be done by a "smart boy, or lazy farmer, or old man" without difficulty. By the end of the First World War, though seven- and

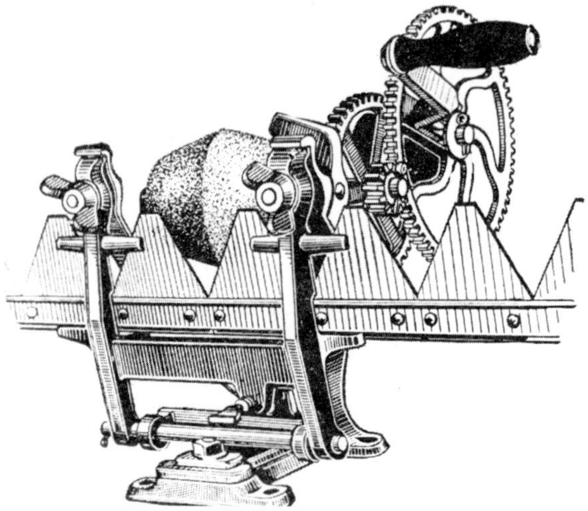

*Left: A specially beveled grindstone was needed to keep the sickle sharp on the mowing machine.*

*Below: In 1853, Lewis Miller became the first inventor to place the seat behind the sickle. This was an important safety precaution. Other inventors and manufacturers adopted this safety feature as shown here on an 1878 model of a Kirby mower. (Smithsonian Institution.)*

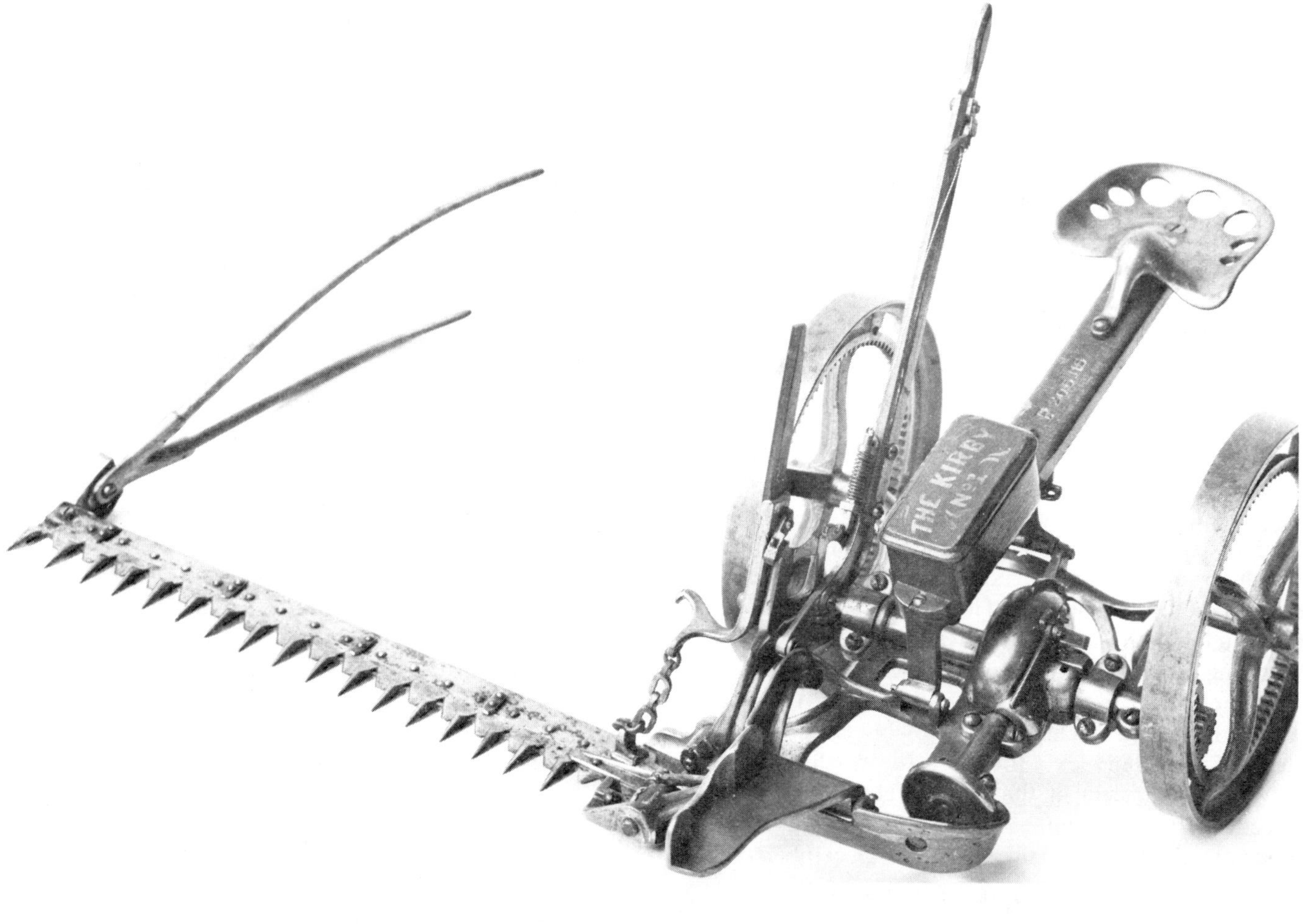

*In 1882, the Bullard Hay Tedder was built by the Richardson Manufacturing Company of Worcester, Massachusetts. By the 1890s, steel had replaced wood for tedder frames. The Quail All Steel Tedder, manufactured by the Ohio Rake Company of Dayton, Ohio, is a notable example.*

eight-foot mowers were becoming popular; these machines cut fifteen acres per day.

The mowing machine required only one tool — a grinder — to maintain maximum operating efficiency. The commonly used flat grindstone, which was turned with a crank, would not fit between the teeth of the sickle blades. Consequently, by the late 1860s, a different type of grindstone was developed. The grindstone had two faces, each set at a sixty degree angle, to fit the bevel of the teeth. A clamp held the sickle against the stone, and, as the hand-crank turned, the stone revolved and ground a sharp edge on the teeth.

## TEDDERS

When the hay was especially heavy, it might not dry properly. If the hay was too damp when it was placed in the stack or in the barn, it might mold or create enough heat to start a fire. Sometimes the hay became packed as the horses and mowers were driven over the previously cut swath as the farmer made his next round. In order to loosen the hay and permit the air to circulate through it and thereby speed the drying process, some farmers fluffed it with a tedder.

The tedder's origin is unknown, but it was invented about 1800, probably by the British. The tedder was not adopted by many American farmers until after the Civil War, when the mowing machine became widely used. The hay tedder had a number of wire prongs or forks attached to a revolving crank shaft which extended between two wheels. As the implement moved forward, the shaft revolved and the prongs kicked backward and lifted the hay. The forks were placed at one foot intervals and most tedders had about eight forks, each composed of several tines. At first, the tedder had a wooden frame, but by the early 1890s, it was being made from steel, and more than fifty makes were on the market.

Tedders worked best about half a day after the hay was cut. If the hay leaves were too dry, though, the tedder would break them off and thereby cause a substantial loss of hay. Since the side-delivery rake lifted and fluffed the hay nearly as well as the tedder, and left it in windrows as well, most eastern and midwestern farmers preferred using it instead of the tedder. By so doing, they saved the expense of purchasing an additional implement. In the Great Plains and Far West, the dry climate cured the hay rapidly and farmers in those regions seldom used the tedder.

## LOADERS

After the hay had been cut and tedded, it was ready for the barn or the stack. The hay sweep eliminated the task of loading hay onto a wagon with a pitchfork, but some farmers preferred to use another implement — the hay loader — for gathering the hay from the swath or windrow. Although patents were granted for hay loaders as early as 1848, this implement did not become successful until the late 1860's. The hay loader attached to the end of a wagon, and as the implement moved forward, a series of forks and endless chains revolved, scooped up the hay, and carried it to the top of the machine where it fell onto the wagon.

The "Douglas' Hay Loading Machine," patented in 1870, was an early model which worked fairly well. This hay loader consisted of eight endless leather belts attached to two rollers. A belt transferred power from the wagon's sprocket wheel to the loader. Steel spikes on the leather belts caught the hay and carried it to the top of the loader where it dropped onto the wagon. But when the wagon was full, the hay loader could not be easily detached or thrown out of gear. Consequently, both the hay wagon and loader were hauled to the barn or to the hay stack. This type of hay loader was a

*Hay loaders, such as these attached behind the wagon and lifted the hay onto the bed. Farm workers generally disliked the loader because it made them work harder and faster to build the wagon stack than did hand-loading with pitch forks.*

cumbersome implement, and it did not pick up all the hay from the ground. In 1875, a more efficient hay loader appeared on the market, when the Keystone Manufacturing Company of Sterling, Illinois, began producing its machine. Like all hay loaders, this implement attached to the wagon.

The most popular hay loaders had either forks or cylinders with tines which lifted the hay and deposited it on an endless inclined apron which carried it up to the wagon. Two men could load two tons of hay onto a wagon with a loader, if the stack was carefully built. The hay loader became popular among farmers in New York, Pennsylvania, Ohio, Indiana, and Iowa. It did not work very well on uneven ground, or in the wind, and because of these disadvantages, the hay loader never achieved widespread popularity. Most farmers preferred to pick up hay from the windrow with pitch forks or with sweep-rakes.

## HAY FORKS

In the East, farmers generally placed their hay in the barn loft. Sometimes as much as ten or twelve tons of hay might be hauled to the barn during the course of a day. If this hay was pitched up into the barn with hand forks, the work was not only hard, but dirty as well, since dust and dried leaves and stems sifted down upon the men in the wagon.

By the late 1840s, some farmers were experimenting with the horse-fork for lifting hay from the wagon into the barn mow. The horse-fork looked much like a large pitchfork. It had a wooden handle five and a half feet long and a wooden head about twenty-eight inches wide to which long steel tines were attached with screws and nuts. A rope was tied to the fork and threaded through a pulley which was attached to the rafter at the peak of the barn. The rope then passed down through another pulley fixed to the barn floor, and from there to the horse. The man working on the wagon drove the fork into the hay, and as the horse walked forward, the fork load of hay was raised to the door of the mow. A man on the wagon held a rope attached to the handle in order to keep the load level. Once the fork reached the top, a worker in the hay loft swung it inside and emptied the load. The horse was then backed up, the fork lowered, and more hay prepared for lifting. The length of the handle, however, made the horse-fork difficult to use under low roofs, but this problem was soon solved by adding a hinge between the head and the handle. The hinge was tripped with a cord by whomever was working on the wagon. As the fork was lowered, the weight of the head caused the hinge to snap back into place ready for another bite of hay. Where one man could unload a ton of hay in thirty minutes by using a pitchfork, one person could unload that same amount in about six minutes by using a horse-fork. Other innovations for lifting hay into the mow followed.

In September 1864, E. L. Walker, a New York inventor, patented a "harpoon" hay fork. This fork worked in a manner similar to a whaler's harpoon. The harpoon fork, either single or double, had tines from twenty-five to thirty-five inches long with folding barbs housed in the shaft hear the point. The fork was driven into the load of hay, and as the horse tightened the rope

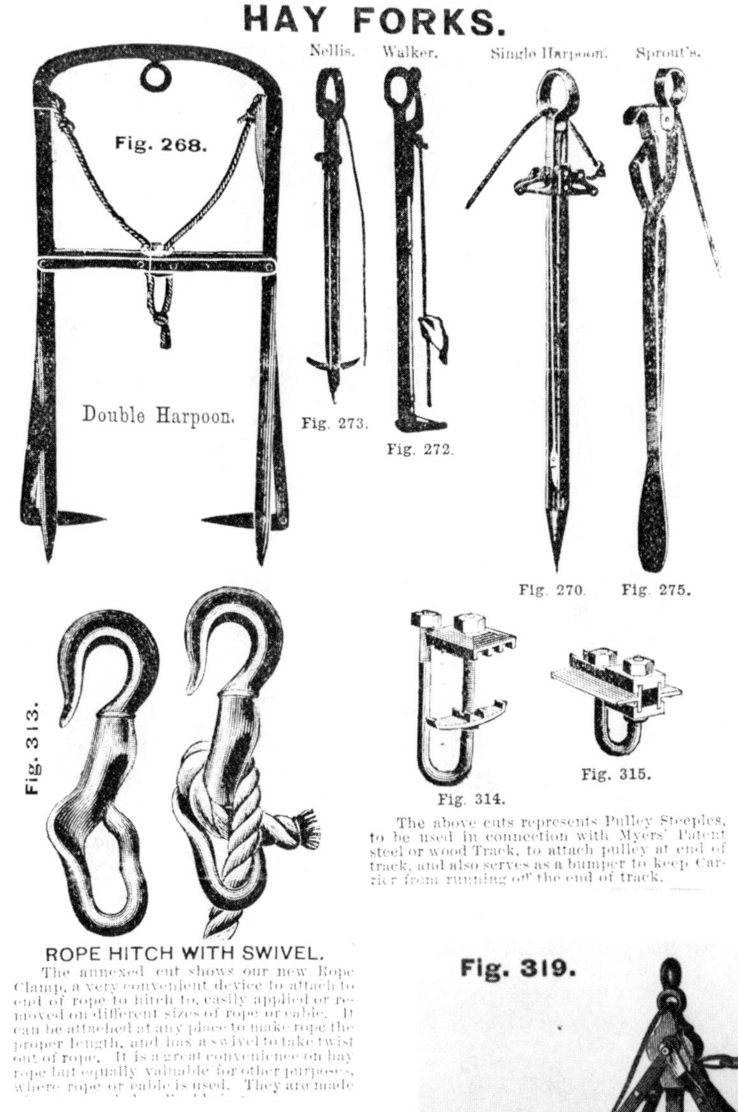


*Harpoon forks were driven into a wagon load of hay in order to lift it into the barn mow. Harpoon forks worked best on long-stemmed hay, but the eagle claw fork worked better on short-stemmed grass or clover.*

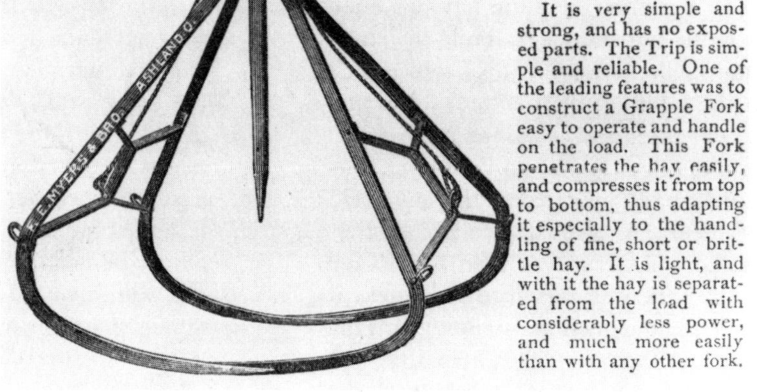


for lifting the fork, the barbs sprung out of the tip and held the hay. The harpoon fork worked best on long-stemmed hay such as timothy. It did not work as well as the horse-fork for short-stemmed hay such as clover, alfalfa, and the short grasses.

About the same time that harpoon forks became popular the "eagle claw" forks were developed. These forks looked like giant bird claws which were hinged together much like a clam shell. The steel teeth were driven into the hay, and, as the horse tightened the slack of the rope, the teeth closed and the load was lifted to the mow. A rope extended from the hinge to the ground, and as the hay was swung into place, the hinge was tripped and the load emptied. In contrast to the harpoon forks, the eagle claw forks worked well on short hay.

*By the late 1860s, farmers were using the steel track and carrier to lift hay into the barn. When the hay fork raised to the track, the carrier rolled to the proper location for release. The release mechanism on the hay fork was tripped by the man on the wagon who held the rope.*

Although hay forks greatly eased the burden of pitching hay up into the barn, once the hay has been released, workers in the loft still had to build the stack by hand. By the late 1860s, however, this problem had been solved with the addition of the steel track and carrier. The track was fixed to the ceiling of the barn and a carrier with a pulley attached to it. A rope was threaded through the pulley and tied to the hay fork. The other end of the rope, passed through another pulley fixed to the barn floor; it was then tied to the horse. When the hay fork was raised to the track, it locked onto the carrier and immediately slid toward the back of the barn. When the hay fork reached the appropriate point for release, someone working in the mow gave a shout to the person holding the trip rope on the ground below. The trip rope was given a jerk, the fork opened, and the hay was released. At that point, the horse was backed up; the fork pulled back along the carrier and lowered with another trip device. Steel hay carriers could support more than half ton of hay and together with the hay fork greatly sped the task of putting hay into the mow.

If a farmer used a sweep-rake to gather the hay and transport it to the barn, he frequently preferred to lift it into the mow with a sling. The sling was essentially a rope net made in two sections and joined together with a trip fastener. The sling was spread on the ground and the sweep-rakes's hay load dropped on it. A horse drawn rope lifted the sling into the barn. The entire load could be raised at once. The sling, however, could be easily pushed out of place when the sweep drove over it, and some farmers preferred to place it on a wagon's empty bed so the load coud be built on top of it. Then, the wagon was simply drawn to the barn and the sling attached to the hoist and lifted into the mow.

## HAY STACKERS

In the midwestern prairie, the Great Plains and the Far West, farmers preferred to stack their hay outside because their annual crops were far too large to fit in most barns. By the early 1860s, some farmers were using tripods made from long poles to build their hay stacks. Two tripods were set up where the farmer plan-

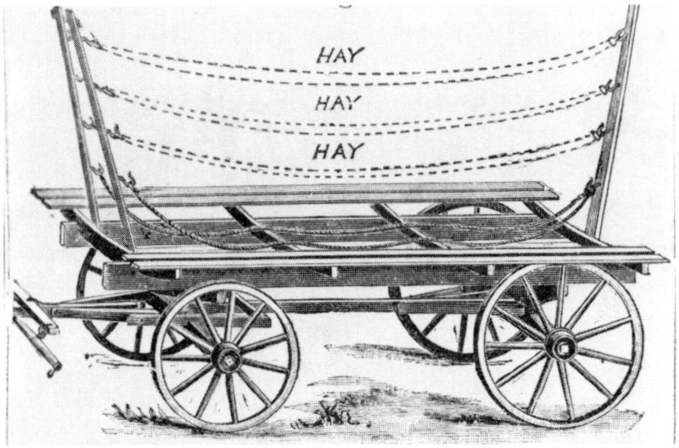

*Some farmers used a sling to lift their hay into the barn mow. In order to do so, they placed a rope sling on the wagon bed and built the stack on top of it. When the wagon reached the barn, the sling was attached to a rope and hoisted into the mow.*

ned to locate the stack. A pulley slid on a rope between each tripod and the hay fork's load was released in the same manner as when it was used in a barn. Two other types of stackers — the overshot and the swinging — soon developed.

The overshot and swinging stackers looked much like sweep-rakes attached to a frame. Hay was piled on the giant fork which was lifted with the aid of a horse, rope, and pulley. As the horse pulled on the rope, the fork lifted up over the frame and tossed the hay onto the stack. The swinging stacker could be locked in place, after the fork was raised, and pivoted over the stack until it reached the proper position for releasing the hay. Some hay stackers such as Palmer's Hay Stacker used a steel horse-fork instead of a large wooden fork to hold the hay. This stacker consisted of a large boom attached to a frame. The fork was sunk into the hay on a wagon and raised until it struck the pulley which automatically swung the fork over the stack for release. Once the hay was dropped, the horse backed up, and, as the fork lowered, its weight caused the derrick to revolve back into position ready for another load. The swinging stacker was particularly useful for loading hay from the stack back onto a wagon for delivery to livestock feed bunks.

## HAY BALERS

During the first half of the nineteenth century, many inventors tried to eliminate the use of pitch forks for stacking hay as well as the nuisance of shipping loose hay by patenting a host of designs for hay presses or balers. The earliest hay presses were designed to work by hand. Hay was stuffed into a box-like cylinder and a hand-crank and wooden plunger squeezed it into a firm bale. These hand presses did not work very well because human strength was insufficient to press the hay adequately, and horse power was soon applied to these implements. By pulling or pushing a lever back and forth or by turning a sweep, a horse could provide the superior strength necessary to compress large amounts of hay into firm, tight bales. In 1853, H. L. Emery of

Fig. 116.—*Stacking Hay.*

*In the Great Plains and Far West, farmers sometimes used tripods with pulleys and hay forks to build the hay stack. Tripods, however, took time to set up and most farmers preferred the overshot or the swinging stacker.*

*Above: The overshot stacker, like this Deering model, lifted the hay and cast it onto the stack.*

*Right: The swinging stacker, like this McCormick implement, raised the hay and swung it over the stack for release.*

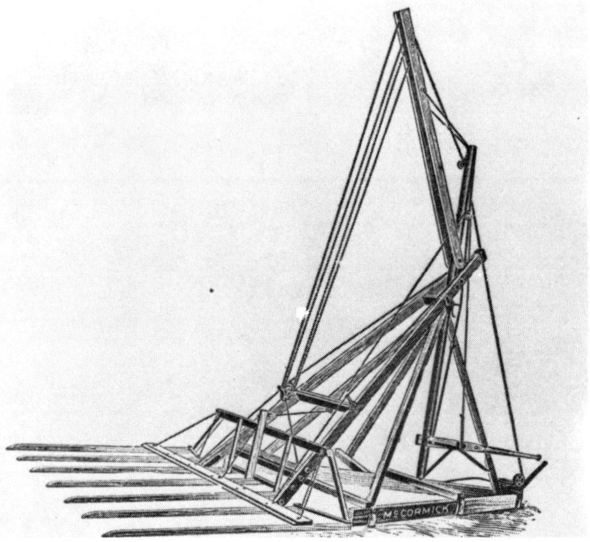


Albany, New York, began manufacturing a horse-powered hay press. With Emery's press, two men and a horse could make five, 250-pound bales per hour. Each bale measured 24x24x48 inches. When the bale was completed wires were fed through slots at each end and the bale tied together. Once the bale was removed, the machine had to be reset to start the baling process all over again.

In 1872, L. & P. K. Dederick also of Albany, New York, began manufacturing a continuous process baler called the "Perpetual Press." This press could form more than one bale at once and the machine did not need to be reset after each bale was completed. As two men fed hay continually into the receiving box of the giant vertical press, horse-powered levers packed the hay tight. When the appropriate bale size had been reached, a block of wood was inserted into the cylinder to mark the end of one bale and to start the beginning of another. The finished bale was then tied with wires and emptied from the machine.

By the mid-1880s, steam-powered balers were on the market. These required a crew of eight men — one engineer and a water hauler to operate the steam engine, two men to pitch the loose hay from the wagon, another to fork it into the baler, two men to tie wires lengthwise on each bale and one man to carry the bale away from the machine. A crew such as this could bale approximately twenty-five tons of hay per day, although some steam-powered presses baled as much as ninety tons per day. Still, horse-powered balers con-

DEDERICK'S NEW HAY-PRESS

DEDERICK'S HAY-PRESS

The close packing of hay for market has become an important object. One of the best machines for this purpose is DEDERICK'S new press, as shown in the accompanying cut. It is a toggle or progressive power press,—its chief improvement being in the parallel arrangement of the toggles, which are so placed that one is near one end, and the other near the other end of the follower; and its movement cannot have any other friction to overcome than that of the material it is pressing. The bulk and weight of the machine are so greatly reduced that the farmer can now have the hay press transported to him as conveniently as by its use he can transport his hay to market. It is worked by a horse upon a capstan—requiring him to make only 5 or 6 circles on a 7 foot sweep to make a bale. With 2 men and a horse, it will bale from 5 to 8 tons of hay per day. Price $100 to $175, according to size. Made by WM. DEERING & Co. of Albany.

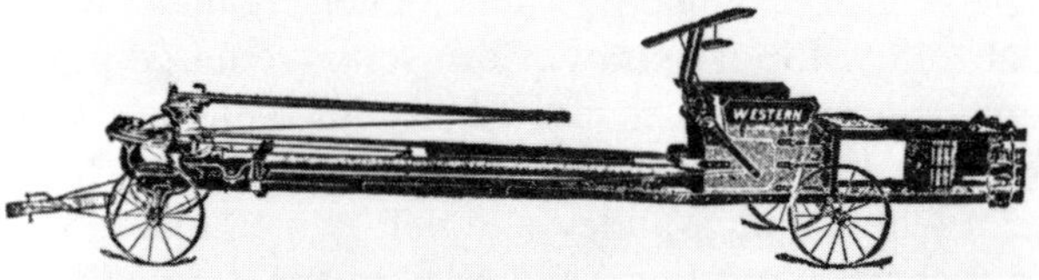


*By the late nineteenth and early twentieth centuries, horse-powered hay balers had been reduced in size from those built during the 1860s. In order to use these smaller presses, the farmer placed the hay in the compression box and the bale was squeezed out of the end at the right. Implements, such as this one, could bale about 18 tons of hay per day.*

*By the mid-1880s, steam-powered hay balers could easily bale 90 tons of hay each day.*

tinued to be made until after the turn of the twentieth century even though these implements only produced about eighteen tons per day. Whether horse-powered or steam-powered, hay balers not only enabled a farmer to stack more hay in a smaller space than ever before, but baled hay also could be easily loaded onto wagons or railway cars for transport and sale. Bales were particularly useful in the prairie hay and alfalfa fields of the Great Plains and Far West where the large quantity of hay was often expensive to hand-stack with hired labor.

## FODDER-CHOPPERS AND FEED MILLS

Since antiquity, farmers have prepared coarse grass and grain stalks for livestock feed by cutting or chopping it into small pieces. Chopped forage was more palatable to cattle and they wasted less of it. American farmers followed this ancient practice, and by the early nineteenth century, they were using hand-powered fodder-choppers to prepare cattle feed.

At first, fodder-choppers simply consisted of a feed box mounted on a wooden or iron frame with a large, lever-like knife attached to one side. Straw or stalks were placed in the box and cut into the desired lengths or pieces. These fodder-choppers could be easily made by the farmer with the aid of the local blacksmith, who fashioned the cutter from an old scythe. By the early

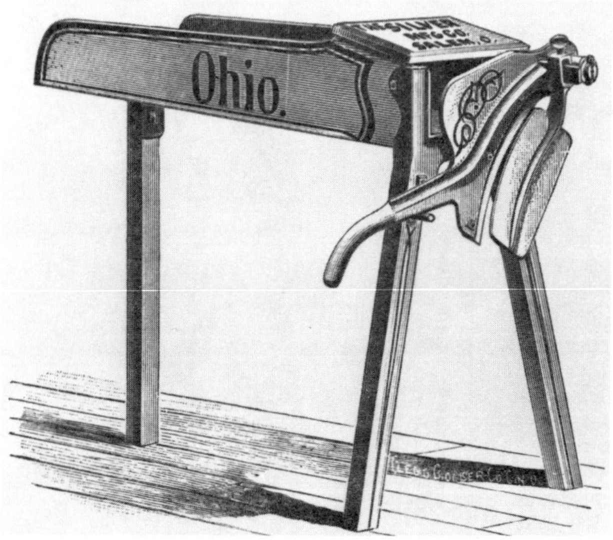

*Fodder-choppers were used to cut corn and other forage crops into small pieces. These lever and crank models were built by the Silver Manufacturing Company of Salem, Ohio, for the 1895 season.*

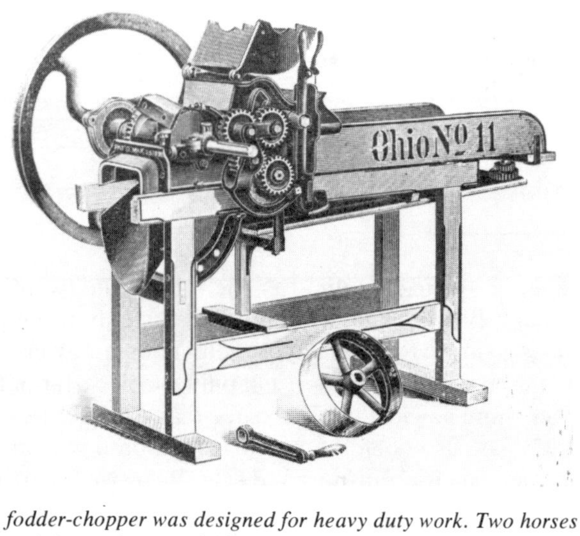


*This fodder-chopper was designed for heavy duty work. Two horses powered this implement which could chop three to four tons of ensilage per hour.*

ROSS No. 108. LITTLE GIANT SELF FEED CUTTER.

1820s, fodder-choppers with spiral knives attached to a shaft revolved as a hand-crank turned. The operator fed the stalks under the knives which cut the forage into pieces. Some fodder-cutters had knives attached to the inside of a flywheel. The straw fed from the box into the spinning wheel. Other models had spiked rollers which fed the crop into revolving knives attached to a spindle. With all models, the chopped fodder fell from the knives onto the ground, ready for feeding livestock.

The lever, crank and flywheel fodder-choppers were suitable only for dry straw. Corn stalks required more chopping power than an individual could generate with a lever or crank for a sustained period of time. However, by attaching a pulley and belt, which linked the fodder chopper to a horse-powered sweep or treadmill, corn stalks could be cut easily. Later in the nineteenth century, steam engines replaced horses for powering large fodder-choppers.

During the last quarter of the nineteenth century, some farmers began storing green or partially cured corn as well as alfalfa and grain sorghum in silos. In order to fill their silos, they used fodder-choppers, known as silage or ensilage cutters, which had been adapted for cutting green stalks. An endless belt fed the crop into radial or spiral cutters. Radial knives were attached to a flywheel and the stalks fed through it. Spiral knives were attached to a spindle and cut the forage much like a reel-type lawn mower. A chain or pneumatic conveyor carried the chopped ensilage into the silo. Before the age of the gasoline tractor, ensilage

*Left: The Little Giant Feed Cutter and Elevator was manufactured by the E.W. Ross Company of Springfield, Ohio. With the aid of a steam engine, this implement could cut eight to twelve tons of ensilage per hour.*

# PATENT SCREW PROPELLER,

OR

# IMPROVED CYLINDRICAL STRAW CUTTER.

## Awarded by the Maryland State Agricultural Society, Nov. 1, 1855, the 1st Premium.

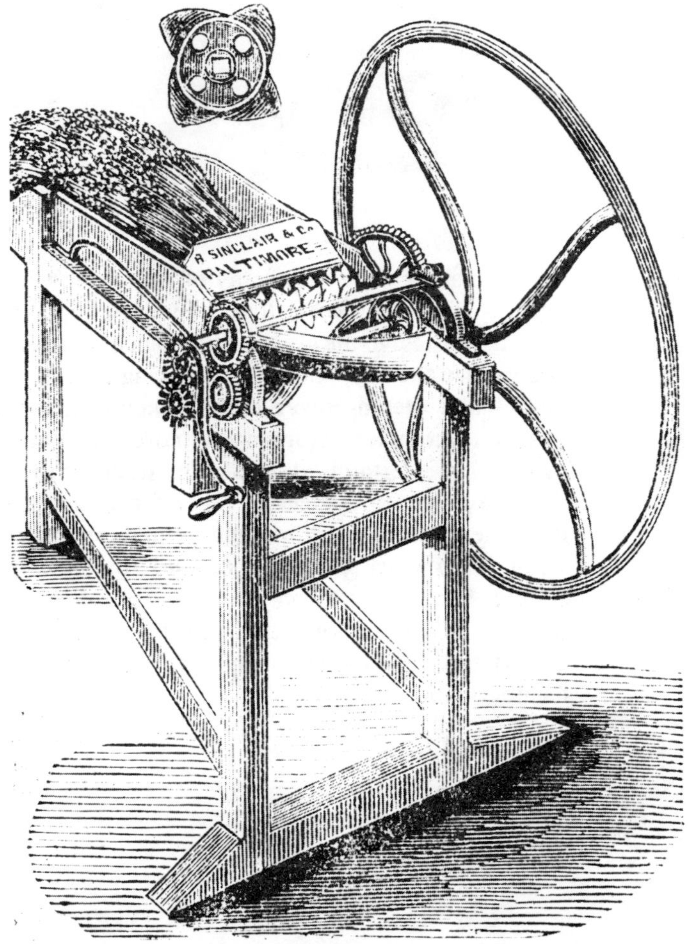


THE CYLINDRICAL STRAW CUTTER, although invented and made in this city more than thirty years past, still retains its high reputation for efficiency and durability. With the present feed works, (Screw or Spiral,) the article is greatly improved and rendered more simple. All efforts of inventors that have been made in Europe or the United States towards improving or inventing Straw Cutters, have failed to produce a machine of equal perfection as our *patent propeller.*

The screwfeed renders the machine simple in arrangement of the works, reduces power, and allows those driven by horse power to be run 30 per cent. faster than ordinary, and producing a consequent greater number of bushels of cut fodder in a given time. The sizes, prices and capacity, are as follows, viz:

| | | | | |
|---|---|---|---|---|
| Width of Cut, | 9. | 11.* | 13. | 15 in. |
| Price, | 28. | 30. | 40. | $45. |
| Capacity per hour, | 45. | 55. | 225. | 250 Bush. |

As regards capacity, the 9 and 11 inches, are rated by hand power, and the 13 and 15 inches by horse.

Also for sale, a variety of Straw Cutters of various sorts and sizes, prices from 5 to $16.

*Manufactured and for sale by*

**R. SINCLAIR, Jr. & Co.**
62 Light street, Balt.

*See figure. Ap. 1.

*Straw cutter advertisement from **The American Farmer-Advertizer**, June, 1860.*

cutters were steam-powered. Only steam engines provided sufficient power to turn the cutting blades and mechanisms which lifted or forced the fine pieces of ensilage into the silo. At the turn of the twentieth century, ensilage cutters chopped about one ton of forage for each horsepower of the steam engine.

After the mid-nineteenth century, many farmers also owned feed mills, which they used for grinding corn, cobs, and small grains into livestock feed. Hand-crank mills appeared first, followed by horse and steam-powered mills. Feed mills worked on the same basic principle of fodder choppers. A feed box dropped the grain or cobs into a chamber where iron mill plates, called burrs, turned and crushed it. If cobs were fed into the mill, a revolving cutter chopped them into fine pieces before the burrs ground the forage into even finer pieces. The burrs were corrugated to carry the crushed feed from the surface of the plates to the outer edge. The milled feed then dropped out of the mill through slots at the bottom. During the 1860s, feed mills, driven by a two-horse sweep or by a railway, crushed from five to ten bushels of corn per hour. By the turn of the twentieth century, horse-sweep or stream-powered mills ground as much as twenty bushels of shelled corn, fifteen bushels of ear corn, or twelve bushels of oats per hour.

In retrospect, no other agricultural activity had a greater variety of tools and implements than did hay and

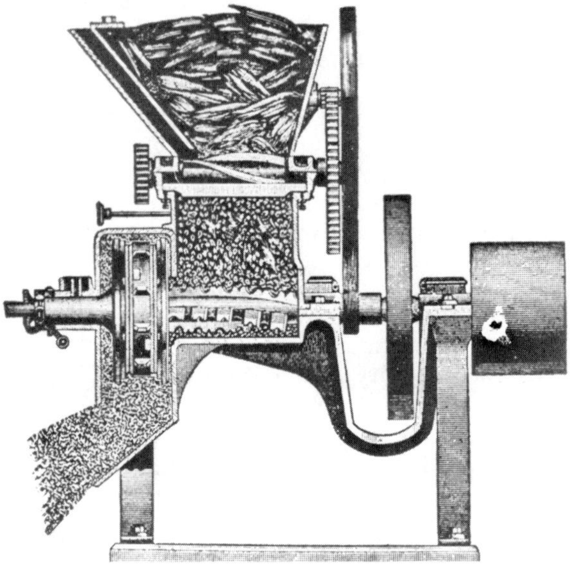

Most farmers who owned livestock and poultry used feed mills to grind grain. Feed mills crushed grain and cobs for easier mixing and feeding. These mills could be operated with either horse or steam power.

fodder-making. The vast array of rakes, mowers, tedders, loaders, stackers, forks, fodder-choppers, and feed mills attest to the great technological change which occurred in the hay- and fodder-making process. These horse- and steam-powered implements enabled the farmer to complete more work more efficiently than ever before. Certainly, the gasoline tractor brought increased power and speeded these farm chores as old implements were modified and as new ones were developed to work properly behind this new power source. But, before the gasoline tractor revolutionized agriculture in the twentieth century, steam-power caught the imagination of agricultural inventors and farmers alike.

*For those farmers who could afford one, a stationary steam engine made many farm chores easier. Washing machines, fodder-choppers, threshing machines and fanning mills could be driven by steam power.*

# CHAPTER X

# STEAM POWER

The exact date is unknown when the first steam engine was used for farm power in the United States. As early as 1807 or 1808, however, Oliver Evans was building steam engines in his Philadelphia shops for powering cotton, flax and wool machinery and for driving sugar cane mills. By 1809, steam engines were turning gristmills, and during the War of 1812, some Georgia planters were using steam to run their cotton gins. Indeed, early in the nineteenth century, southern planters most commonly used steam engines for agricultural purposes, and by the late 1820s, Louisiana plantation owners were making large investments in steam-powered sugar mills. A decade later, only Pennsylvania had more steam engines in operation than did Louisiana. By the outbreak of the Civil War, steam power had almost completely replaced horses or oxen for powering sugar mills, and plantation owners across the South were using steam engines to operate rice mills and to turn cotton gins.

The work which a steam engine performed was a significant increase over that accomplished by horse or man power. With it, for example, three men and a cotton gin could remove the seeds from 1,000 to 4,500 pounds of cotton per day. This was about 100 times more than they could gin without steam power. Until the 1830s, most of the steam engines used in American agriculture were British imports. During the 1830s, however, American manufacturers began making steam engines for agriculturists. The W. Tift and Company of Cincinnati, the West Point Foundry of New York, Holmes Hinckley of Boston, and John Allaire of New York were several of the most important early steam manufacturers in the United States. A decade later, during the 1840s, the Tredegar Iron Works in Richmond, Virginia, also became an important supplier of steam engines, when it began furnishing steam-powered sugar mills to southern planters.

Prior to the 1850s, stationary rather than portable or traction steam engines generated steam power. Stationary engines, as the term implies, could not be moved easily. Rather, these steam engines were bolted to solid, brick and mortar foundations. A belt linked the steam engine to whatever machinery was to be driven. Southern plantations, which were devoted to the inten-

*The portable steam engine reached the market during the 1850s. These steam engines, however, were not self-propelled, and a team of horses had to pull these implements to the threshing site. In 1877, this steam engine was built by the C. & G. Cooper Company in Mount Vernon, Ohio. The smokestack folded back whenever the engine was moved from place to place. The belt, which linked the steam engine to the threshing machine, was attached to the flywheel on the opposite side.*

*In 1886, the Aultman-Taylor Company of Mansfield, Ohio, placed the "Baby Elephant" on the market. It had a 6-horsepower engine, and it could be purchased with skids instead of wheels, if the engine was to be used on a semi-permanent location.*

sive cultivation of specific crops such as cotton or sugar cane, were much better suited for the adoption of stationary steam engines than were farms devoted to the extensive cultivation of various crops. Steam engines did not become popular on grain farms until portable models reached the market. The daily chores on most farms were just too diverse for the convenient use of stationary steam engines. Once the boiler, steam cylinder, and flywheel were mounted on a wheeled frame, the portable (horse-drawn), steam engine became practical for farms concerned with general agriculture. The portable steam engine could be used on any part of the farm no matter whether it was needed in the corn field, the wheat field, outside the barn, or down the road at the neighbor's place.

Although Thomas Jefferson had called for small, portable steam engines as early as 1815, more than three decades passed before that innovation became a reality. In 1849, A. L. Archambault of Philadelphia manufactured the first American-made steam engine which had the advantage of mobility. Archambault aptly named his steam engine the "Forty-Niner," and built it in 4-, 10-, and 30-horsepower models. By the outbreak of the Civil War several dozen agricultural manufacturing companies also were building portable steam engines all of which were designed for belt work, that is, to power farm implements such as threshing machines, corn shellers, and fanning mills.

Although the portable steam engine appeared on the market in increasing numbers during the 1850s, and although farmers recognized its labor saving value, steam power was not used on a widespread agricultural basis prior to the Civil War. At best, before 1860, the application of steam to general agriculture was still very much in the experimental stage. During the 1850s, steam engines were still primarily found either on southern plantations or on the large farms of wealthy land owners in the midwestern prairies. Few farmers had even seen a steam engine and fewer still owned one, but most believed this innovation would only supplement the role of the draft horse and mule rather than replace them on the farm.

During the 1860s and 1870s, however, important technical changes in plowing, planting and harvesting machinery helped to increase grain production. With expansion in production, additional technical changes were needed to enlarge the threshing machine's capacity and to increase its operating speed. As manufacturers built larger threshing machines, more power was needed to operate them. The horse-powered treadmills and sweeps simply were not capable of adequately turning the mechanisms of the large threshing machines. Moreover, horses seldom maintained the even speed required for maximum threshing efficiency. If the horses walked too fast, all of the grain might not be threshed out of the heads. If the horses walked too slowly, poor separation took place. Both problems meant either wasted or dirty grain and economic loss for the farmer. Furthermore, horses had to be rested or changed frequently.

Because of these problems, by the late 1860s, an increasing number of grain farmers were beginning to use portable steam engines to power their threshing machines. By 1868, for example, the Wood and Mann Steam Engine Company of Utica, New York, had sold

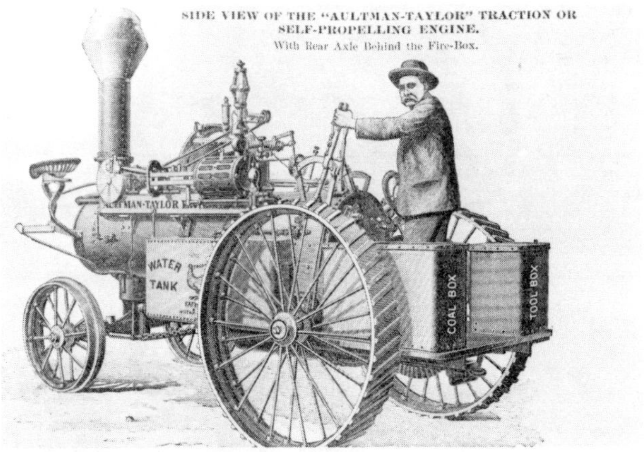


*In 1885, the Aultman-Taylor Company placed this 15-horsepower traction steam engine on the market. Two horses, however, still had to be hitched in front to help guide the engine down the road. Notice the seat in front.*

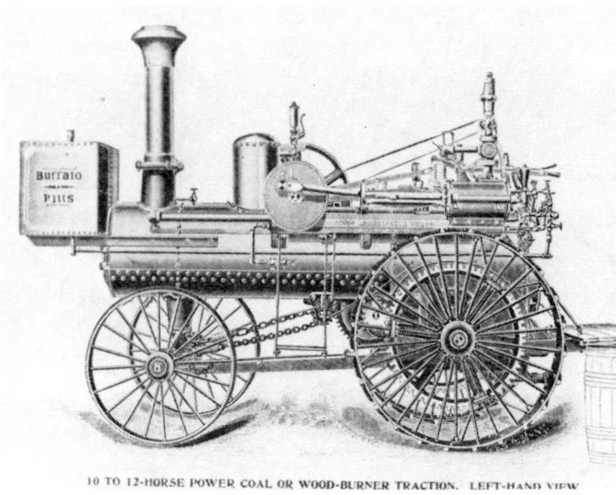


*In 1894, the Pitts Agricultural Works in Buffalo, New York, built this steam engine. It was designed to power threshing machines, and to burn wood or coal.*

more than 1,200 steam engines. At that time, a 12-horsepowered engine could drive a threshing machine capable of threshing more han 100 bushels per hour. These engines used a half a cord of wood or from 300 to 400 pounds of coal during a ten-hour day. By 1870, the building of steam engines was an important aspect of the threshing machine manufacturer's business. During that decade, farmers in all sections of the nation began applying steam power to their wheat, corn and rice threshing and cotton ginning operations. By 1880, the Census Bureau estimated that steam power threshed 80 percent of the grain harvest in the major wheat producing states. Most of those steam engines were manufactured by the various threshing machine companies, Indeed, itinerant or custom threshermen often bought their steam engine and threshing machine as one outfit.

Although the portable steam engine increased the speed and capability of the threshing machine, it had one serious flaw — the portable steam engines were not self-propelled. Lacking traction, steam engines could neither plow nor move about under their own power. Consequently, their use was limited to belt work on threshing, ginding, milling, and ginning machines. In order to move the portable steam engine, the operator had to hitch a team of horses to it and pull it from place to place. Some steam engines had a seat or platform on the front or the rear from which the operator steered it with the reins. On other models, the farmer simply held the reins as he walked alongside the engine.

Although self-propelled steam engines did not reach the market before the 1870's, it was not because inventors failed to devote their attention to the development of traction models. As early as 1769, Nicholas Cugnot, a French inventor, built a self-propelled steam engine which he drove through the streets of Paris. But, the first traction steam engine produced in the United States was not marketed until 1873, when the Battle Creek, Michigan, firm of Merritt and Kellogg offered a self-propelled model for sale. By the late 1870s, the C. and G. Cooper Company of Mount Vernon, Ohio, had won the reputation of being the first company to manufacture traction steam engines in quantity and market them nationwide. In 1880, more than 1,000 Cooper steam tractors were in use across the country. Other agricultural manufacturing firms quickly sought to win the farmer's business, and, by 1881, most of the threshing machine manufacturers were making steam traction engines. By the late 1880s, the gearing, clutching, braking, and steering problems had been eliminated so that the driver could operate the tractor without losing control.

All steam tractors essentially shared a common design, no matter which company built them. Each had a boiler, engine (cylinder, piston, and valves), governor, flywheel, traction gears, wheels and a firebox. The heat in the firebox converted the water in the boiler to steam by passing down a flue. A valve admitted the steam into the cylinder where it drove the piston joined by the connecting rod to the crankshaft. In turn, the crankshaft moved the gearing linked to the rear wheels. Traction steam engines were made from two possible designs. One design called for a frame upon which the boiler, engine, and other parts were mounted. The second design involved making the boiler the main structure and attaching the various parts to it. Usually, the engine was attached to the boiler and the boiler then mounted on a frame or truck.

Traction steam engines were built with either a direct-flue or a return-flue boiler. The direct or locomotive-flue was virtually universal for traction steam engines intended primarily for plowing. With this flue, the heat and smoke passed from the firebox in the rear

*The Frick steam engine (ca. 1875) needed a team of horses to pull it from place to place. It did not have a seat, however, and the driver walked alongside and guided the horses with the reins. (Smithsonian Institution.)*

through flues or tubes to the front of the boiler where it exited up through the smokestack. As the heat and smoke moved forward, it heated the water which surrounded the flues and turned the water into steam. In the return-flue boiler the heat and smoke moved forward through a large flue then passed up and back down the boiler in several smaller flues from which it exited through a smokestack at the rear of the boiler. The return-flue boiler had the best fuel economy, since the heat passed through the boiler twice, but the smokestack at the rear increased the heat around the engineer. Some operators argued, however, that the direct-flue boiler was stronger and safer than the return flue boiler.

Indeed, a major problem with the early traction steam engines was that the boiler iron was sometimes too weak to withstand the pressure generated inside. When water converts to steam at $212^{o}$ Fahrenheit, it expands 1,600 times in volume; adequate safety valves were needed to prevent explosions and the inevitable injury of the operator and others nearby with flying pieces of shrapnel and scalding water. By the late 1870s, however, manufacturers were using Bessemer steel and improved joint-making techniques for their boilers and the hazards of steam engines were greatly reduced.

Sparks escaping from the smokestack created another steam engine hazard, particularly when threshing was done in the stubble field. The addition of spark arrestors, made by placing screens in the smoke-stack or by forcing steam into it, helped reduce that danger, but cautious threshermen kept the steam engine as far away from the straw stack as possible. An equally serious danger involved the weight of the traction steam engine. Bridges designed for horses, wagons, and carriages could seldom suport a ten- or twenty-ton steam engine. Only improved bridge engineering eliminated this problem, but changes came slowly and many steam engines crashed through the timbers and into creeks and ravines before improvements were made.

Although steam tractors only lumbered along at a rate of two or three miles-per-hour, the sight of these iron monsters, with smoke spewing from their stacks, invariably terrified oncoming horses. After numerous incidences of bucking horses, overturned wagons, and runaway teams, many states passed laws requiring that a team of horses be hitched in front of the steam engine while it traveled down a public road. The horses in front tended to give reassurance to oncoming teams that all was well and the number of accidents decreased.

By the late 1870s, farmers were beginning to use self-propelled steam engines in considerable numbers; and, by the mid-1880s, the steam engine met the grain farmer's major threshing needs. Many agriculturists, however, wanted the steam tractor to do more than propel itself from field to field and thresh grain, shell corn, gin cotton, or power a wood saw. In the vast expanse of the Great Plains and Far West, wheat farmers anxiously awaited the development of a steam engine that could pull the plow.

Prior to the development of the steam threshing engine, inventors had been working for many years on a plowing engine. No one knows when the first steam plowing engine was made or by whom, although it may have been Luke Johnson, a Leominster, Massachusetts inventor, in 1816. Other inventors grappled with the problem of producing an engine that had sufficient power to propel itself across a field and pull a plow at the same time. During the 1830s, 1840s, and 1850s, various attempts were made to build a steam plow, but, at best, most inventors achieved only limited success.

Edmund C. Bellinger of South Carolina was one such inventor. Bellinger believed a steam engine should not waste its power moving across a field. His steam engine, patented on 19 November 1833 had a cable and windlass attachment. Bellinger's technique was to place the steam engine at the side of the field and use its power to draw a gang of plows, attached to a cable, back and forth across it. This arrangement allowed the en-

Russell Traction Engine
SIX, TEN, THIRTEEN AND SIXTEEN HORSE

*The Russell & Company of Massillon, Ohio, built this traction steam engine during the 1890s. (Smithsonian Institution.)*

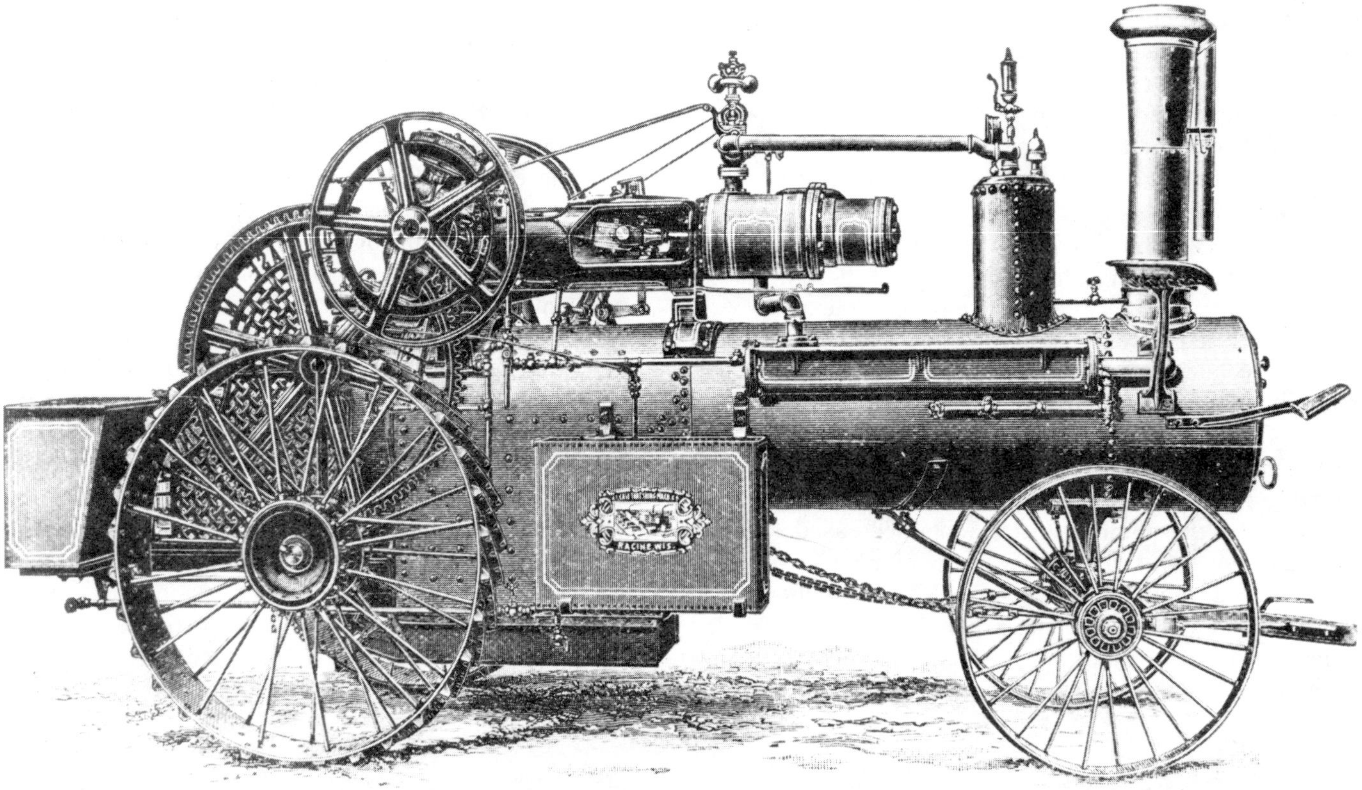


*The J.I. Case Company of Racine, Wisconsin, built compound cylinder steam engines during the early twentieth century. Compound cylinders provided the extra power needed for operating large threshing machines. (Smithsonian Institution.)*

*During the late nineteenth and early twentieth centuries, the Huber and Buffalo Pitts Companies built steam engines which burned straw for fuel. These steam engines were only suitable for threshing, since the straw burned rapidly and a constant supply from the stack was necessary. Both engines have return flue boilers.*

*Left: This front view of a Huber steam traction engine shows the reserve water tank swung to the side and the flues through which the heat and smoke traveled.*

*Below: Steam threshing in Kansas, ca. 1910-1915. Notice the wind stacker on the threshing machine. (Kansas State Historical Society.)*

gine to move ahead slowly as the plowing progressed, yet at the same time devote all of its power to pulling the cable and plow attachment.

Some inventors tried to modify Bellinger's steam plowing method by using two steam engines, one on each side of the field, to draw the plow back and forth. Bellinger's steam plow, however, never won widespread support in the United States. Two major faults prevented its success. First, the cost of a steam engine with plow, cable and windlass apparatus was extremely high. Only the most wealthy farmers could afford an outlay of from $10,000 to $12,000 at mid-point of the nineteenth century. Secondly, cable plowing with steam power was impractical on American wheat farms. Here, the fields were too large for a cable apparatus to work efficiently. At least four men were needed to operate the cable plow, and they could plow only about twenty acres per day. The fields of Great Britain were much smaller and much more suitable for this method of steam plowing, but in the Great Plains and Far West, the fields were too large. There, farmers needed a steam engine which could pull a gang of plows behind it easily, and which required only one operator and a fuel and water tender. Until steam engines were built strong enough to pull a gang of plows, the grain farmers of the prairies and the plains preferred using horse power for plowing their fields.

Obed Hussey was perhaps the most famous early inventor to build a steam plow. In 1855, Hussey tested a steam plow before the Maryland Agricultural Society, and, in 1856, he exhibited his steam engine at the Indiana State Fair. Although Hussey won a silver cup in the Indiana competition he was unable to gain the financial backing needed to proceed with his experiments, and he abandoned further work on the steam plow. If Hussey was the most famous inventor to work on the

*Steam threshing in Russell County, Kansas, 1908. (Smithsonian Institution.)*

*Steam threshing in Russell County, Kansas, 1909. (Smithsonian Institution.)*

*Steam threshing scene ca. 1908-1909. Notice the grain chute emptying from the threshing machine into the wagon. A water tender is alongside the steam engine at the right. (Smithsonian Institution.)*

*Steam plowing in Meade County, Kansas. Notice the disk plow and the corrugated clod-crusher. (National Archives.)*

ing outfit of John W. Fawkes, a Lancaster, Pennsylvania, inventor, probably attracted the most public attention prior to the perfection of a commercial steam plowing engine.

Fawkes' steam engine, exhibited at the Illinois State Fair in 1858, weighed ten tons, extended eighteen feet long and eight feet wide, and had a coal box mounted above the front wheels. A vertical boiler held 360 gallons of water and two nine-inch pistons with a fifteen-inch stroke generated 30-horsepower. The piston rods were geared to a roller 6 feet long and 6 feet in diameter. This roller served as the drive wheel. The two forward wheels were linked to the steering column, and the operator steered the tractor from a platform beside the boiler. A frame, attached to the rear of the steam engine, held eight fourteen-inch plows. The depth of each plow could be adjusted and the entire gang lifted from the soil with a lever. Fawkes' steam plow moved at a rate of three or four miles-per-hour, and cut a furrow nearly nine and a half feet wide, and required an 18-foot radius to turn effectively. Its operating costs were approximately $2.50 per acre. Even so, horse-drawn plows could till an acre for about half that amount.

In that same year, 1858, Thomas H. Burridge of St. Louis, Missouri, also invented a steam traction engine designed for plowing. Burridge's implement, patented on 31 July 1860, consisted of two steam engines or pistons mounted on an iron frame at each side of a vertical boiler. Two drive wheels ten feet in diameter propelled the tractor and a gang of plows attached to the frame at the rear. Like Fawkes' steam plow, Burridge's implement was unwieldly and it did not come into general use.

Ten years after Fawkes tested his steam plow in Illinois, on 10 May 1868, P. H. Standish, an inventor at Pacheco near Martinez, California, patented a vastly different steam plowing apparatus called the "Mayflower." This steam plow had a vertical boiler and two horizontal engines which generated 12-horsepower. At the rear of the steam tractor two to four vertical shafts were attached to a series of rotary diggers or tillers which were geared to the engine. Each digger was made from six knives. The diggers revolved horizontally on a perpendicular shaft and tore or stirred the soil as the tractor moved forward, much on the principle of the modern rotary lawn mower. The tillers cut a twelve foot furrow from two to six inches deep. Like Fawkes' steam plow before it, the Standish invention was a huge, clumsy machine. It extended twenty-four feet in length and twelve feet in width. The two rear drive wheels were eight feet in diameter and thirty-two inches wide. The front wheel was a mere four feet in diameter and eighteen inches wide. Weighing eight tons, the Standish rotary steam plow had a forward speed of from 1.7 to 3.4 miles-per-hour and it could till five acres an hour.

Traction steam engines, such as the Fawkes, Burridge, and Standish, remained ineffective until the 1870s, because inventors had great difficulty working out an efficient power-weight ratio. Because of the steam tractor's great weight, most of the power was used to propel it across the field or down the road. Until

*This Avery steam traction engine is pulling a ten-bottom plow. The engine has been mounted on the frame beneath the boiler. Implements with this design were called "undermounted" steam engines. (National Archives.)*

*In 1908, this 60-horsepower "Big Betsy" steam engine, built by the Holt Company, seeded grain in the San Joaquin Valley on ground too soft for horses. Each wheel was 7½ feet wide, and the two outer wheels on each side could be removed before the engine traveled between fields. This steam engine was nearly 50 feet wide and too cumbersome to be practical. (National Archives.)*

*This 1905 Holt steam traction engine pulled a combine which cut at a rate of 65 acres per day. The roller in front of the steam engine helped smooth the ground and keep the engine level. (National Archives.)*

farmers preferred to use horse power for plowing rather than to invest in an expensive "elephantine" iron monster.

Several technical problems involving power and weight had to be overcome before a fully satisfactory steam engine could be used for plowing. First, the cast-iron gearing of the steam engines designed for threshing were only strong enough to move the tractor from one place to another. Cast-iron gears could not withstand the strain placed upon them during drawbar work. Second, in order to gain sufficient traction, inventors at first, utilized the engine's weight instead of an efficient combination of gears. Although traction could be improved by placing most of the weight above the rear wheels, these steam engines were usually so heavy they bogged down in the field, particularly in damp or soft soil. Lastly, the plows, harrows, and grain drills on the market were designed for horse power. Horse-drawn implements did not work properly behind steam tractors. Consequently, plowing, seeding and harvesting operations would not be improved until new implements were developed to work efficiently with the increased draft of a traction steam plowing engine.

By the late 1870s, manufacturers were making great strides toward the development of a steam tractor which could pull a plow as well as operate a threshing machine. Bessemer steel strengthened the gearing which meant that greater power could be applied to the traction wheels and to the plows behind. Differential gearing and friction clutches made the steam engine more maneuverable than ever before. Improved gearing gave steam tractors two forward speeds — fast and slow. Compound or double cylinders were added to use the steam more efficiently and to provide more power than did single cylinder models. Manufacturers also increased fuel and water capacities. Since a 10-horsepower engine burned about one and a half cords of wood per day and used about 700 gallons of water, adequate carrying capacity was important to shorten refueling time and to eliminate as much fuel and water hauling as possible. Steel boilers made high pressures possible and improved steam valves prevented loss of valuable power. Manufacturers tried to capitalize on these improvements by giving their steam engines names which suggested power, speed, and unequalled performance, such as the Robinson "Conqueror," the Harrison "Jumbo," the Minnesota "Little Giant," the Advance "Incomparable," the Frick "Eclipse," the Monitor "Champion," the Port Huron "Rusher," the Geiser "Peerless," and the Northwest "New Giant."

Even though production of steam engines increased rapidly during the last quarter of the nineteenth century, the average farmer was still hesitant to make the large investment required for ownership of a steam tractor. Consequently, itinerant or custom threshermen owned most of the steam engines used to power threshing machines. But, farmers in the Great Plains and Far West were beginning to make the investments

*In the Great Plains and Far West, steam engines had to plow as well as power threshing machines, because of the large fields and harvests. In 1909, this 14-bottom Peerless steam-lift plow was at work on a farm near Fullerton, North Dakota. (Courtesy Hiram M. Drache.)*

necessary to buy steam engines which could be used to capacity for threshing and plowing during most of the year. By the 1890s, monstrous steam tractors, some weighing twenty-five tons with 120-horsepower, moved relentlessly across the wheat lands of the American West. These traction steam engines easily plowed from thirty-five to forty-five acres per day. The largest of those steam engines pulled twenty to thirty plows and tilled as much as seventy-five acres per day and decreased plowing costs from two dollars to forty cents per acre. By the early twentieth century, 110-horsepower steam engines in the Far West and Great Plains simultaneously pulled plows, grain drills, and harrows. These outfits covered a strip as much as thirty feet wide at a rate of three or four miles-per-hour and covered from eighty to one hundred acres per day. By so doing, these steam traction engines replaced forty to fifty teams of horses with accompanying implements and men.

At the turn of the twentieth century, implement companies were building more than 5,000 traction steam engines annually — an increase of 3,000 tractors over the previous decade — with the J. I. Case and Huber Companies leading the competition. The most powerful steam engines designed primarily for plowing usually had two cylinders. A two-cylinder steam engine could start heavy loads easily without damage to the gears and provided better balance than did single piston machines. These steam engines were also "double acting," that is, steam was admitted into the main cylinder in an alternating sequence. When the proper amount of steam was in the cylinder, a valve shut off the intake, and the piston moved to the end of the cylinder. When the piston reached the end of its stroke, the steam was released. At the moment of release, another valve admitted more steam to drive the piston back to the other end of the cylinder. Not all of the steam was released each time so that some would cushion the piston head at the end of each stroke. Each piston was fitted with two or three rings to prevent the steam from passing between it and the cylinder wall and thereby cause a loss of power.

By 1900, steam tractors were also fitted with governors which regulated forward speed. Coil springs had been placed between the boiler and the front and rear wheels and under the steering gear to prevent breakage and to help cushion the tractor from rough spots in the fields or on the roads. The fuel supply also became more flexible as tractors were designed to burn coal, wood, oil, and straw. Although the use of straw helped reduce the hay stack during threshing time and eliminated the need to procure other fuel, it was not convenient when the steam engine was used for plowing. Straw burned too quickly and large amounts had to be hauled continually to the tractor in the field. These improvements did not mean that the traction steam engine had been made smaller and lighter. In fact, most tractors remained gigantic and weighed from ten to twenty-five tons

In retrospect, through the nineteenth and early twentieth centuries, steam engines, whether stationary, portable, or traction, replaced thousands of draft animals on the American farm. By 1913, steam engines provided power equivalent to 7,000,000 horses and mules. Still, most farmers could not afford to own a steam engine. From 1908-1915, during the peak of the steam engine's popularity, only one farmer in twenty owned a steam tractor. Where steam power was used for threshing and plowing, most farmers usually hired custom operators to do the work for them. By the beginning of the First World War, however, the gasoline tractor was rapidly replacing the steam engine for threshing and plowing. With the International Harvester Company's production of the first affordable, row crop tractor in 1924 — the Farmall — the age of steam was relegated to the past.

# *APPENDIX*

## Metallurgy and Technological Change in American Agriculture

From the age of hand power through the age of steam, rapid technological change in agriculture would have been impossible without similar change in metallurgy, that is, in the art and science of making iron and steel. Indeed, metallurgy has always been closely linked to the manufacture of farm tools. Early in the colonial period, Americans recognized the importance of making iron and steel in order to provide themselves with a wide variety of agricultural tools such as axes for clearing the land, plow shares for tilling the soil, hoes for cultivating crops, and scythes and sickles for reaping the harvest.

As early as 1619, the Virginia Company of London granted a group of Southampton entrepreneurs the right to build an ironworks in the vicinity of Jamestown for the purpose of smelting, casting and forging iron. In 1622, however, hostile Indians killed the workers, destroyed the ironworks, and temporarily ended all attempts to produce iron in Virginia. By the mid-1640s, however, the American iron industry had been successfully established in eastern Massachusetts, and the Bay Colony became the leader in colonial iron production until the early eighteenth century.

The ironworks of colonial America supplied local blacksmiths with the metal necessary for making agricultural tools. During the seventeenth century, for example, village blacksmiths crafted farm tools from bloomery iron. Ironmasters made bloomery iron by placing ore and charcoal into a Catalan forge. The charcoal burned with the aid of an air bellows and reduced the ore to metal. This forge, however, could not produce sufficient heat to liquify the iron. Instead, the iron formed a semi-molten lump or bloom in the bottom of the forge. Each bloom contained impurities, called slag, which the iron workers removed by pounding the metal with a water-powered trip hammer. This hammering or forging also increased the density of the metal and shaped it into a bar. Bloomery iron was reheated and hammered several times until it was ready for the blacksmith to fashion into plow shares, scythes, and hay forks.

Occasionally, farm tools were made at the ironworks instead of at a blacksmith's shop. In 1647, for example, Joseph Jenks began making scythes from bloomery iron at the Lynn Iron Works in Massachusetts. Eight years later, the General Court granted him a patent for making an improved scythe. This scythe was lighter and easier to maneuver than the English scythes which most American farmers used, because the blade was longer, thinner and stronger. Jenks' scythe eased the New England farmer's task of mowing hay, and it was the most important agricultural tool produced at this early American ironworks.

Bloomeries, such as the Lynn Iron Works, were scattered throughout many of the colonies, particularly in Massachusetts, New Jersey and Pennsylvania. Ironmasters with bloomery forges, however, could not produce large quantities of wrought-iron. At best, they could only meet the needs of local blacksmiths. Consequently, increased demands for iron necessitated improvements in smelting techniques. Early in the eighteenth century, smelting became more advanced when ironmasters began installing blast furnances in order to meet the demands of a growing market. The blast furnance was much larger than the bloomery forge, and it generated temperatures high enough to melt iron. As a result, colonial ironworks began producing more and better iron for the making of agricultural tools.

The blast furnance was a large, hollow, brick structure which held alternate layers of iron ore, charcoal, and limestone. A water-powered bellows provided oxygen for the burning charcoal and sped the smelting process. As the ore melted, the limestone helped separate the impurities from the molten metal which ran to the bottom of the furnance; the slag floated to the top. The molten iron was drained or tapped off by removing a plug at the bottom of the furnance. As the liquid iron ran out through a trough, it collected in sand molds sculpted in the ironworks' floor. There, it solidified into long bars of cast-iron called "pigs," which contained from three to five percent carbon and other impurities. Generally, two tons of iron ore smelted into one ton of pig iron.

When the pig iron cooled, workers carried it to a nearby forge for refining. At the forge, the pig iron was reheated in a charcoal fire and worked into a lump called a "half-boom." Then, a water-powered hammer pounded the carbon and other impurities from the metal until the iron acquired the tough, fibrous structure of wrought-iron. Small forges usually produced about two tons of wrought-iron or bar iron per week, but large forges, with several hearths and hammers, turned out more than 300 tons annually — a far greater amount than a bloomery was capable of producing. Blacksmiths and ironworkers used this bar iron, just as they had used bloomery iron, to fashion plow shares, scythes, hoes, shovels, axes, and wagon tires. During the late eighteenth century, for example, New York foundrymen made good quality scythes, hoes, and spades; and, a small manufacturer in Berkshire, Massachusetts, made 1,100 rakes annually from blast furnace iron.

Blast furnace iron not intended for the blacksmith

was cast immediately into agricultural tools and household utensils. In 1797, Charles Newbold cast the first plow in the United States and other inventors made similar experiments. At the turn of the nineteenth century, Peter Townsend, who operated the Sterling Iron Works in Orange County, New Jersey, cast three plows that were "no heavier than the old fashioned" wooden implements. He also attempted to cast fanning mills and corn pickers of some sort, but without success. About that same time, Peter J. Curtenius, a New York City foundryman, also advertised cast-iron plows for sale.

Virtually all inventors at that time had great difficulty using cast-iron, because metallurgists could not regulate consistently the amount of carbon left in smelted iron. Since high percentages of carbon gave cast-iron a crystalline structure, which caused tools made from it to break easily, most cast-iron was used for making hollow ware such as kettles, skillets, and pots, or for making stove plates and Dutch ovens. During the late 1820s and early 1830s, though, metallurgists learned how to restructure the carbon content of pig iron by remelting it, thereby making it strong enough to cast into agricultural tools. By 1830, a Massachusetts foundry cast more than 1,000 plows annually; and, a Pittsburgh factory produced 100 cast-iron plows per day. By 1850, the art of plow-making had been transferred from the local blacksmith to ironworks which specialized in plow casting primarily because of improved techniques for making cast-iron. The blacksmith no longer had to pound out plow shares from wrought-iron or plate wooden moldboards with thin iron strips. Instead, farmers now ordered standardized cast-iron plows directly from the ironworks or from implement manufacturers.

Other scientific and technical changes improved early nineteenth century metallurgy. Many ironworks, for example, adopted the puddling process for refining pig iron. The puddling process, first used in 1816 at the Plumsock Rolling Mill in Fayette County, Pennsylvania, involved melting pig iron in a reverberatory furnace. This furnace held the iron and fuel in separate chambers. The heat from the firebox melted the pig iron, and the carbon oxidized or burned away as workers stirred the molten metal and exposed it to the air. As the carbon burned out, the molten iron lost its fluidity and formed a pasty bloom or lump of nearly pure wrought-iron. Rollers flattened the bloom into bars which were ready for sale to merchants and blacksmiths.

The puddling process for making wrought-iron had two major advantages over making wrought-iron in an open forge. First, greater quantities of pig iron could be refined in a reverberatory furnace than could be heated and hammered out at an open forge. Second, since the iron was kept separate from the fuel, coal could be used instead of charcoal because the sulphur and phosphorus in that carbon fuel would not transfer to the molten iron.

The adoption of the coal-fired reverberatory furnace brought great change to the American iron industry. By the 1830s, coal was replacing charcoal for fuel in the furnaces of eastern Pennsylvania. Change came slowly, however, because wrought-iron made from charcoal was tougher and more malleable and had a better welding quality than did iron smelted from coal. Nevertheless, anthracite, and later bituminous coal, was used on an increasing basis, because it was less expensive than charcoal. With the increased use of coal for fuel, the iron industry began to concentrate west of the Appalachians. There, ironmasters capitalized on the abundant coal deposits as well as on the vast iron ore ranges in the vicinity of the Great Lakes. By the end of the Civil War, coal was the primary fuel for making iron.

In spite of the importance of the iron industry for the innovation and manufacture of agricultural tools, steel production expanded very slowly in the United States. Indeed, throughout the colonial and early national period of American history, pig iron and wrought-iron satisfied most needs for agricultural tools. Technical problems as well as high costs for transportation, and fuel and labor shortages kept steel production low until the mid-nineteenth century. As early as 1655, however, a Long Island ironmaster made steel; and, by 1750, a steel furnace was in operation in Kinningworth, Massachusetts. During the latter half of the eighteenth century, steel furnaces were also in operation in New York and Pennsylvania.

Early American steel was called "blister" or "cemented" steel. Blister steel was wrought-iron with enough carbon bonded on the surface to make steel suitable for manufacturing edge-tools. Ironmasters made blister steel by placing wrought-iron bars, with alternate layers of powdered charcoal dust, in a furnace. The furnace was kept at a cherry-red heat for a week or more, during which time the iron absorbed the critical amount of carbon from the charcoal to form steel. Ironmasters called this steel-making method the "cementation" process. During the cementation process, a chemical reaction occurred between the carbon of the charcoal and the oxygen and slag trapped in the wrought-iron which caused blisters to form on the surface of the metal. When, the bar cooled, it was ready for market.

Blister steel-making was a slow, expensive process, because the ore first had to be reduced to wrought-iron then reheated with great amounts of charcoal to give the metal the appropriate carbon content. By 1810, only 917 tons of blister steel were produced annually in the United States. Blacksmiths used it to make plow shares, to plate hoes, and to make edge-tools. By 1820, one implement manufacturer in New Haven, Connecticut, used blister steel to make pitchforks.

In 1812, the quality of American steel began to improve when a Valley Forge ironmaster began making crucible steel. The crucible steel process, involved placing wrought-iron and powdered charcoal into a crucible or container above a furnace flame. As the iron melted, it absorbed about one percent of the carbon and became steel. This molten steel was then cast into ingots and rolled into slabs ready for making various agricultural tools and implements. By the 1830s, Pittsburgh manufacturers were producing cast steel hoes, shovels, and hay- and manure-forks which were superior to similar tools made in Europe. In 1846, the Pittsburgh firm of Jones & Quigg rolled the first slab of plow steel cast in the United States for John Deere of Moline, Illinois. This plow steel was low grade, however, and Pittsburgh furnaces did not begin to make high quality cast steel for agricultural tools until 1853. Even then, the total production remained low. By 1860, thirteen crucible steel plants produced less than 12,000 tons of high grade steel for agricultural implements, springs, and locomotive wheels. Brittle spots and blow holes or gas bubbles and inadequate crucible materials kept early nineteenth century ironmasters from making the best crucible steel possible. By the late nineteenth century, the use of graphite instead of silica crucibles improved the quality of the steel. Still, the crucible process could not produce great quantities of steel for the multiplicity of industries, including the agricultural implement industry, in need of it.

By the mid-nineteenth century, changes in agricultural technology created demands for greater steel production. Inventors and manufacturers required steel for plow shares and for harvesting machine components. These demands were soon met with a new steel-making technique called the Bessemer process. This process was developed in 1855 by Henry Bessemer, an English inventor. It enabled ironmasters to convert large amounts of molten pig iron into steel. Other metallurgists improved the Bessemer process, but it essentially involved forcing a blast of air into a large container, called a converter, which held molten pig iron. The oxygen in the air burned away the carbon, silica, and manganese from the pig iron. Since some carbon was needed to give the metal strength, iron with a high carbon and manganese content, called spiegeleisen, or a compound of iron, manganese, and carbon, called ferro-manganese, was added to properly carbonize the metal, and thereby, turn it into steel. The liquid steel was then poured into molds and rolled into blooms for sale to various manufacturers and railroad companies.

The first Bessemer steel made in the United States was produced in Wyandotte, Michigan, in 1864. During the remainder of the nineteenth century, it was primarily used for making rails and, thereby, helped expand the railroad industry. Some Bessemer steel was also used for moldboard plows and for the gearing of steam traction engines. The Bessemer process, however, did not remove the sulfur and phosphorus from the iron. Therefore, only ores which did not contain those elements could be smelted and converted into steel.

In 1868, however, the New Jersey Steel and Iron Company at Trenton began making steel with the open-hearth process. This method involved using natural or bituminous gas fuel and a preheated air blast to superheat a reverberatory furnace to as much as 4,000 degrees Fahrenheit so the impurities would burn away from the molten pig iron. Since pig iron melts at about 2,100 degrees and wrought-iron liquifies at about 2,500 degrees, the open-hearth furnace could easily keep steel, which melts at about 2,800 degrees Fahrenheit, in a molten state. Consequently, even though the open-hearth process was slower than the Bessemer method, which took only about ten to fifteen minutes for conversion, the steel could be kept in a liquid state longer while metallurgists experimented with the proper carbon content to make various grades of steel. Frequently, open-hearth furnaces were added to Bessemer works to utilize scrap steel and to remelt worn-out rails. Open-hearth steel was particularly well suited for making boilers and fireboxes for steam engines.

By the turn of the twentieth century, open-hearth steel was a major competitor of cast-iron; and, in 1908, it surpassed Bessemer steel in volume produced. Open-hearth furnaces could be economically operated with a monthly production of 1,000 tons, whereas Bessemer plants had to produce 8,000 to 10,000 tons to warrant the financial investment. At a time when the United States Steel Corporation was attempting to monopolize the market and to keep prices high, the open-hearth process enabled many small, independent producers to stay in business.

By 1870, the iron and steel industry clearly was concentrated west of the Appalachians and Pittsburgh was the center of the industry. This concentration reflected more than new sources of iron ore and coal or improved methods for making iron and steel. It also meant that the traditional relationship between the local ironworks and the village blacksmith had been altered for all time. Large quantities of iron and steel were sold directly to manufacturers such as McCormick, Aultman-Taylor, and Deering, who used it to make agricultural implements. In 1902, the newly created International Harvester Company even built its own steel plant and rolling mill near Chicago to satisfy its metal needs. As a result of this change, the local blacksmith no longer used his forge to craft farms tools from bars of wrought-iron. Instead, he used it to repair the iron and steel implements made elsewhere.

During the late nineteenth century, increased iron and steel production enabled implement manufacturers to expand their operations. The dramatic growth in the value of farm machinery in the United States reflects the increased production of agricultural implements.

Between 1850 and 1870, for example, the value of farm machinery increased from $152,000,000 to $271,000,000. In 1900, it rose to $750,000,000; and, ten years later it reached $1,265,000,000. This growth in value reflects, in part, the importance of the advances in metallurgy. Progress in metallurgy enabled inventors and manufacturers to design and to build efficient, affordable iron and steel implements. Furthermore, the iron and steel parts of plows, cultivators, harvesters, threshing machines, combines, hay-making equipment, and steam engines hastened the completion of the farmers' tasks and eased his labors.

Certainly, without technical progress in the iron and steel industry, technical change in the manufacture of agricultural tools and implements would have been impossible. Ironmasters worked to meet the farmers' technological needs, and inventors of agricultural implements capitalized on the progress made in metallurgy. By so doing, each contributed to the technological changes in agriculture which helped the American farmer meet the food needs of a nation that was becoming increasingly industrialized and urbanized.

## SUGGESTED READINGS

For a detailed study of metallurgy and the iron and steel industry, see the following suggested readings:

*Appleton's Cyclopedia of Applied Mechanics*. 2 vols. New York: D. Appleton and Company, 1893.

Boylston, H.M. *An Introduction to the Metallurgy of Iron and Steel*. New York: John Wiley & Sons, 1928.

Bining, Arthur Cecil. *Pennsylvania Iron Manufacture in the Eighteenth Century*. Vol. IV. Harrisburg: Publications of the Pennsylvania Historical Commission, 1938.

Bishop, Philip W. "The Beginnings of Cheap Steel," U.S. National Museum, Bulletin 218, *Contributions from the Museum of History and Technology*, Washington, D.C., 1959, pp. 27-48.

Byrn, Edward. *The Progress of Invention in the Nineteenth Century*. New York: Munn & Co., 1900.

Clark, Victor S. *History of Manufacture in the United States, 1893-1928*. Vol. III. New York: Peter Smith, 1949.

Hunter, Louis C. "Heavy Industries Before 1860." In *The Growth of the American Economy*, pp. 210-228. Edited by Harold F. Williamson. New York: Prentice-Hall, Inc. 1944.

_____ . "Influence of the Market Upon Technique in the Iron Industry in Western Pennsylvania Up to 1860," *Journal of Economic and Business History I* (February, 1929): 241-81.

_____ . "The Heavy Industries Since 1860." In *The Growth of the American Economy*, pp. 467-95. Edited by Harold F. Williamson. New York: Prentice Hall, Inc., 1944.

*Iron and Steel: Principles of Manufacture, Structure, Composition and Treatment*. Machinery's Reference Series No. 36. New York: The Industrial Press, 1909.

Kaempffert, Waldemar. *A Popular History of Invention*. 2 Vols. New York: Charles Scribner's Sons, 1924.

Kranzberg, Melvin, and Pursell, Carrol W. (eds.). *Technology in Western Civilization*. Vol. I. New York, Oxford, University Press, 1967.

Lewis, W. David. *Iron and Steel in America*. Greenville, Delaware: The Hagley Museum, 1976.

Saunders, Clyde A., and Gould, Dudley C. *History Cast in Metal*. N.P.: Cast Metals Institute, American Foundrymen's Society, 1976.

Stoughton, Bradley. *The Metallurgy of Iron and Steel*. New York: McGraw-Hill, 1934.

Swank, James M. *History of the Manufacture of Iron in All Ages*. Philadelphia: The American Iron and Steel Association, 1892.

# BIBLIOGRAPHY


Ardrey, P. L. *American Agricultural Implements.* Chicago: The Author, 1894.

Anderson, Russell H. "Grain Drills Through Thirty-Nine Centuries," *Agricultural History* 10 (October, 1936): 157-205.

*Appleton's Cyclopedia of Applied Mechanics.* 2 Vols. New York: D. Appleton and Co., 1893.

Bailey, Liberty Hyde. *Cyclopedia of American Agriculture.* 4 Vols. New York: Macmillan Co., 1909.

Bidwell, Percy Wells and Falconer, John I. *History of Agriculture in the Northern United States, 1620–1860.* New York: Peter Smith, 1941.

Bourne, Patricia M. and Sageser, A. Bower. "Background Notes on the Bourne Lister Cultivator," *Kansas Historical Quarterly* 22 (August, 1952): 183-86.

Byrn, Edward W. *The Progress of Invention in the Nineteenth Century.* New York: Munn & Co., 1900.

Carpenter, Edward. "The Groundhog Thresher: An Enigma," *Wisconsin Magazine of History* 37 (Summer, 1954): 217-18.

Cathey, Cornelius O. "Agricultural Implements in North Carolina, 1783-1860," *Agricultural History* 25 (July, 1951): 128-35.

Chase, Lew Allen. "Hiram Moore and the Invention of the Harvester," *Michigan Historical Magazine* 13 (July, 1929): 501-05.

Church, Lillian. "History of Corn Planters," United States Department of Agriculture, Information Series No. 69, 1935.

_____. "History of Cultivators," United States Department of Agriculture, Information Series No. 52, 1935.

_____. "History of Grain Drills," United States Department of Agriculture, Information Series No. 70, 1935.

_____. "Partial History of the Development of Grain Harvesting Equipment," United States Department of Agriculture, Information Series No. 72, 1939.

_____. "Partial History of the Development of Grain Threshing Implements and Machines," United States Department of Agriculture, Information Series No. 73, 1939.

Clymer, Floyd. *Album of Historical Steam Traction Engines and Threshing Equipment.* Los Angles: The Author, 1949.

Cooper, Martin R., Barton, Glen T., and Brodell, Albert P. "Progress in Farm Mechanization," United States Department of Agriculture, *Miscellaneous Publication No. 630,* 1947.

Danhof, Clarence H. *Change in Agriculture: The Northern United States, 1820-1870* Cambridge: Harvard University Press, 1969.

_____. "Gathering the Grass," *Agricultural History* 30 (October, 1956): 169-73.

_____. "The Tools and Implements of Agriculture," *Agricultural History* 46 (October, 1972): 81-90.

David, Paul A. "The Mechanization of Reaping in the Ante-Bellum Midwest." In *Industrialization in Two Systems: Essays in Honor of Alexander Gerschenkron,* pp. 3-39. Edited by Henry Rosovsky. New York: John Wiley & Sons, 1966.

Drache, Hiram M. *The Day of the Bonanza.* Fargo, North Dakota: Institute for Regional Studies, 1964.

Ellis, L.W. and Rumley, Edward A. *Power and the Plow.* New York: Doubleday, Page & Co., 1911.

Faulkner, Harold Underwood. "Farm Machinery and the Industrial Revolution," *Current History* 33 (March, 1931): 872-76.

Fishwick, Marshall. "Sheaves of Golden Grain," *American Heritage* 7 (October, 1956): 80-85.

Gray, Lewis Cecil. *History of Agriculture in the Southern United States to 1860.* 2 Vols. Glouster, Massachusetts: Peter Smith, 1958.

Giedion, Siegfried. *Mechanization Takes Command.* New York: Oxford University Press, 1948.

Gray. R. B. *The Agricultural Tractor, 1855–1950.* St. Joseph, Michigan: American Society of Agricultural Engineers, 1975.

Greeno, Follet L. *Obed Hussey.* Rochester, New York: The Author, 1912.

Hardeman, Nicholas P. *Shucks, Shocks, and Hominy Blocks: Corn As A Way of Life In Pioneer America.* Baton Rouge: Louisiana State University Press, 1981.

Hedrick, Ellwood. *Lewis Miller.* New York: G. P. Putnam's Sons, 1925.

Heilman, Grant. "Threshing Season," *American West* 11 (November, 1974): 20-29.

Higgins, F. Hal. "The Celebrated Steam Plow of Philander H. Standish," *American West* 3 (Spring, 1966): 24-27, 91.

_____. "The Combine Parade," *Farm Quarterly* 4 (Summer, 1949): 42-47, 94, 96, 98, 100, 102, 104.

_____. "The Moore-Hascall Harvester Centennial," *Michigan Historical Magazine* 14 (Summer, 1952): 415-37.

Hill, Charles L. "The First Combine," *Wisconsin Magazine of History* 35 (Summer, 1952): 263-66.

"History of the Plow," United States Department of Agriculture, Bureau of Agricultural Engineering, *Information Series No. 48,* n.d.

Hutchinson, William T. *Cyrus Hall McCormick.* 2 Vols. New York: D. Appleton-Century Co., 1930, 1935.

_____. "The Reaper Industry in Midwestern Agriculture, 1855-75." In *Essays in Honor of William E. Dodd,* pp. 115-30. Edited by Avery Craven.

Chicago: University of Illinois Press, 1935.
Johnson, Paul C. *Farm Inventions in the Making of America.* Des Moines, Iowa: Wallace-Homestead Book Co., 1976.
Jones, C. Clyde. "Letters of Philander H. Standish," *Agricultural History* 31 (April, 1957): 8-18.
Jones, E. R. "A Bit of Twine Knotter History," *Wisconsin Magazine of History* 12 (December, 1928): 225-27.
Jones, Robert Leslie. "The Introduction of Farm Machinery into Ohio Prior to 1865," *The Ohio State Archaeological and Historical Quarterly* 58 (January, 1958): 1-20.
Kellar, Herbert A. "Agricultural Machinery and Westward Expansion," *The Westerners Brand Book* 4 (September, 1947): 37-44.
_____. "The Reaper as a Factor in the Development of the Agriculture of Illinois, 1834-1865," *Illinois State Historical Society Transactions* 34 (1927): 105-14.
Kendall, Edward C. "John Deere's Steel Plow," U. S. National Museum, Bulletin No. 218, *Contributions from the Museum of History and Technology.* Washington, D.C., 1959, pp. 15-25.
Kohlmeyer, Fred W. and Herum, Floyd L. "Science and Engineering in Agriculture: A Historical Perspective," *Technology and Culture* 2 (Fall, 1961): 368-80.
Lewton, F. L. "Notes on the Old Plows in the United States Museum," *Agricultural History* 17 (January, 1943): 62-64.
McClure, H. B. "Haymaking," United States Department of Agriculture," *Farmers' Bulletin No. 943,* 1918.
McCormick, Cyrus. *The Century of the Reaper.* Boston: Houghton, Mifflin Co., 1931.
McKinley, Marvin. *Wheels of Farm Progress.* St. Joseph, Michigan: American Society of Agricultural Engineers, 1980.
Miller, Merritt Finley. "The Evolution of Reaping Machines," United States Department of Agriculture, *Farmers' Bulletin No. 103,* 1902.
Norbeck, Jack. *Encyclopedia of American Steam Traction Engines.* Sarasota, Florida: Crestling Publishing, 1976.
Nyberg, Chris. "Highlights in the Development of the Combine," *Agricultural Engineering* 38 (July, 1957): 528-29, 535.
Olmstead, Alan L. "A Reappraisal of the Mechanization of Reaping and Mowing in American Agriculture, 1833-1870,,' *Journal of Economic History* 35 (June, 1975): 327-52.
Pursell, Carroll W. Jr. *Early Steam Traction Engines in America.* Washington, D.C.: Smithsonian Institution Press, 1969.
Quick, Graeme R. and Buchele, Wesley F. *The Grain Harvesters.* St. Joseph, Michigan: American Society of Agricultural Engineers, 1978.
Rasmussen, Wayne D. "The Civil War: A Catalyst of Agricultural Revolution," *Agricultural History* 39 (October, 1965): 187-95.
_____. "The Impact of Technological Change on American Agriculture," *Journal of Economic History* 20 (December, 1962): 578-91.
_____. "The Mechanization of Agriculture." In *Agricultural Literature: Proud Heritage Future Promise, A Bicentennial Symposium, September 24-26, 1975*, pp. 295-315. Edited by Alan Fusoni and Leila Moran. Washington, D.C. Associates of the National Agricultural Library and the Graduate School Press, United States Department of Agriculture, 1977.
Rogin, Leo. *The Introduction of Farm Machinery in Its Relation to the Productivity of Labor in Agriculture of the United States During the Nineteenth Century.* Berkeley: University of California Press, 1931.
Ross, Earle D. "Retardation in Farm Technology Before the Power Age," *Agricultural History* 30 (January, 1956): 11-18.
Schafer, Joseph. "Hiram Moore, Michigan-Wisconsin Inventor," *Wisconsin Magazine of History* 15 (December, 1931): 234-43.
Schlebecker, John T. "The Combine Made in Stockton," *The Pacific Historian* 10 (Autumn, 1966): 14-21.
_____. *Whereby We Thrive: A History of American Farming, 1607-1972.* Ames: Iowa State University Press, 1975.
Spence, Clark C. "Experiments in American Steam Cultivation,,' *Agricultural History* 33 (July, 1959): 107-16.
_____. "The 'Lancaster': An Early Steam Plow," *Pennsylvania History* 25 (October, 1958): 271-83.
Steward, John F. *The Reaper.* New York: Greenberg, Inc., 1931.
Swingle, Frank B. "The Invention of the Twine Binder," *Wisconsin Magazine of History* 10 (September, 1926): 35-41.
Thomas, John J. *Farm Implements and Farm Machinery.* New York: Orange Judd & Co., 1869.
Thwaites, Reuben Gold. "Cyrus Hall McCormick and the Reaper," *Wisconsin State Historical Society Proceedings* 56 (1908): 234-59.
Whitehead, Vivian B. *A List of References for the History of Agricultural Technology.* Davis, California: Agricultural History Center, University of California, 1979.
Wik, Reynold M. *Steam Power on the American Farm.* Philadelphia: University of Pennsylvania Press, 1953.
Williams, Mentor L. "Cooper and the Moore-Hascall Harvesting Machine," *Michigan History* 31 (March, 1947): 26-34.
Williamson, Harold F. *The Growth of the American Economy.* New York: Prentice Hall, 1944.
Yerkes, Arnold P. and McClure, H.B. "Harvesting With the Sweep-Rake," United States Department of Agriculture, *Farmers' Bulletin No., 838,* 1917.
Zintheo, C. J. "Corn-Harvesting Machinery," United States Department of Agriculture, *Farmers' Bulletin No. 303* 1907.

# *INDEX*